REINDEER DON'T FLY

EXPLORING THE EVIDENCE-LACKING REALM OF EVOLUTIONARY PHILOSOPHY

MICHAEL EARL REIMER

PATH LIFE TAKES PUBLISHERS

Is the principle of evolution indisputable? Is Darwinism derived from scientific fact or elaborate theory? In his new book *Reindeer Don't Fly*, Michael Earl Riemer explores the history, science, and multitude of misconceptions that form the dogma we call evolution. With pungent humor, pinpoint logic, and theological expertise, the author's lively narratives offer a passionate and compelling argument that upends the lockstep view that evolution dare not be questioned.

. . . the study of evolution is an exercise in futility, for it serves no useful propose. Are you aware that over the course of many decades, the study of evolutionism's doctrines has not produced a single useful irrefutable fact, historical information, or any pertinent scientific finding or advancement that would be of benefit to mankind? Of course, at times a bit of useful information may have come out of its study, but that would be in spite of, not because of.

Chapter 7, *Reindeer Don't Fly*

. . . prebiotic chemistry is a way to determine how processes that never happened, under conditions which never existed, during a time that never was, made it possible to create an organism that never lived, using a synthesis of materials brought into being by nothing.

Chapter 8, *Reindeer Don't Fly*

REINDEER DON'T FLY
Exploring the Evidence-Lacking Realm of Evolutionary Philosophy
Copyright © 2018 by Michael Earl Riemer

Permission is hereby granted to any church, mission, magazine, newsletter, book, or periodical to reprint or quote from any portion of this book on the condition that the passage is quoted in context, and due acknowledgement of the source be given. Any portion of this book may be freely distributed, if it is without charge or obligation. The only qualification for the above permission is that the name of the author and his email address be given, and all references from the quoted portions be provided in full.

Unless otherwise indicated, Bible quotations are taken from the King James Version.

Path Life Takes Publishers
San Francisco, Agusan Del Sur
Philippines

Cover Design: Fran Platt, Eden Graphics, LLC.
Print and Digital Formatting: Dayna Linton, Day Agency

ISBN: 978-1-7329061-0-5 (Print)
ISBN: 978-1-7329061-1-2 (Digital)

Author's Email: eldermike547@yahoo.com

REINDEER DON'T FLY

EXPLORING THE EVIDENCE-LACKING REALM OF **EVOLUTIONARY PHILOSOPHY**

ACKNOWLEDGMENTS

I wish to express my gratitude to the following people who contributed in some way to the writing and publishing of this book.

My thanks to evangelist and author Ralph Woodrow for his encouragement. Director of *Wild Ministries*, Mike Prom, had a number of good suggestions for improving the structure and layout of the book. I offer my sincere appreciation to "sharp as a tack" researcher Michelle Smallback, who reviewed the manuscript and made important contributions in the form of pertinent questions, needed corrections, observations, and sound suggestions. Alexander Doak and Chris McMahan provided encouragement and helpful corrections and suggestions. I appreciate the gracious comments and counsel from Paul Lagan and Bruce Zatkow. Edward E. Stevens from the International Preterist Assoc. made a few pertinent observations and suggestions.

Francine Eden Platt of *Eden Graphics, Inc.* did a great job on the cover design. I also extend my appreciation for editorial support by *Editage* (www.editage.com) for English language editing.

CONTENTS

ABOUT THE AUTHOR xv
PREFACE xvii
INTRODUCTION 1

CHAPTER 1 - HOW OLD IS THE EARTH? 11
    THE BOOK OF GENESIS: LITERAL OR FIGURATIVE HEBREW POETRY? 17
    ADAM, THE FIRST MAN 20
    ARCHBISHOP JAMES USSHER 24
    DAY-AGE THEORY 28
    RUIN-RECONSTRUCTION THEORY (GAP THEORY) 36
    AN UNNECESSARY COMPROMISE 37

CHAPTER 2 - FROG PUREE—THE STUFF OF LIFE 45
    PROPAGANDA IN THE TEXTBOOKS 45
    EVOLUTIONISM—THE STEALTH RELIGION 50
    WE HAVE A BIT OF TIME, SO LET'S GO TO THE LAB AND CREATE LIFE 54
    A BIT MORE COMPLICATED THAN WE THOUGHT 60
    LET LIFE BEGIN 63
    THREE CRUCIAL INGREDIENTS 65
    IRREDUCIBLE COMPLEXITY 66
    THE NON-EXISTENT PROGRAMMER 69

RULE OF METHODOLOGICAL NATURALISM..................72

MACROEVOLUTION AND MICROEVOLUTION 72

CHAPTER 3 - THAT MYSTERIOUS SOMETHING 79

CRYONICS...83

DOCTOR HULBERT 85

CHAPTER 4 - WALKING AMONG HORSE- AND PUMPKIN-KIND 89

LARGE AND SMALL CANINES 92

MINIATURE AND DRAFT BREED HORSES....................93

THE PUMPKIN KIND..97

FELIX THE FLYING FROG: A PARABLE ABOUT SCHEDULES, CYCLE TIMES, AND SHAPING NEW BEHAVIORS..................99

GREAT WHITE SHARKS 101

A FLOOD OF MISINFORMATION 103

MCINTOSH APPLES 106

CHAPTER 5 - TRANSITIONAL FORMS AND COMMON ANCESTORS...109

MUSH ON, SANTA CLAUS 112

SANTA CLAUS: AN ENGINEER'S PERSPECTIVE..........113

UNQUESTIONED BELIEF 115

VESTIGIAL ORGANS 127

NEANDERTHAL AND CRO-MAGNON MAN PRECONCEPTIONS 128

IT'S NOT ROCKET SCIENCE......................................135

CHAPTER 6 - LET'S GO SNIPE HUNTING 147

| INFAMOUS EVOLUTIONARY HOAXES | 149 |
| THE PEPPERED MOTH AND NATURAL SELECTION | 154 |

CHAPTER 7 - JUST ADD WATER AND SOME TIME 161

A PETITE PRIMATE NAMED LUCY	162
A ZIPPERLESS ZIPPER	164
THE INVISIBLE CARETAKER	167
THE BACK TO EDEN METHOD OF ORGANIC GARDENING	170
A WHOLE LOT OF SHAKING GOING ON	176

CHAPTER 8 - BALONEY ON THE MENU 185

| DON'T GET TAKEN FOR A RIDE WITH NASA | 190 |
| BALONEY DETECTION KIT | 193 |

CHAPTER 9 - THERE'S A SUCKER BORN EVERY MINUTE 199

DEADLY HYDROHYDROXIC ACID	200
GOD'S BENEFACTION TO MANKIND	204
PSEUDOSCIENCE	224

CHAPTER 10 - DATING ROCKS AND DEAD THINGS 231

RADIOMETRICITY, THE ART OF DATING ROCKS	234
LIVING FOSSILS	246
FRESH DINOSAUR BONES	248
LET'S TRY AN EXPERIMENT	250
INDEX FOSSILS	253

CHAPTER 11 - EXTRA-LARGE CRITTERS 257

| WHO DELIVERED THE WATER? | 261 |
| BEREFT OF THE GIANTS | 265 |

MODERN ANIMAL GIANTS 272

CHAPTER 12 - GOLIATHS AND PRODIGIES 275

HISTORICAL ACCURACY OF THE OLD TESTAMENT 277

EARLY POST-FLOOD CIVILIZATIONS AND PUZZLING
ANCIENT TECHNOLOGY 278

CIVILIZATION BEFORE THE FLOOD 285

ANCIENT GIANTS 286

MODERN-DAY GIANTS 288

INTELLECTUAL ABILITIES AND QUALITIES OF THE
ANTEDILUVIANS 291

CHAPTER 13 - THE GROUND BENEATH US 299

SEDIMENTARY ROCK 302

FORMATION OF STRATA AND FOSSILS 306

CASE OF THE VANISHING PERPETRATOR 310

WHAT LIES BURIED DEEP BENEATH OUR FEET? 314

CAPTAIN NOAH DISEMBARKS 316

THOUSANDS OF FEET OF DEFORMED SEDIMENTS 318

WHERE DID THE THOUSANDS OF FEET OF SEDIMENTARY
ROCK COME FROM? 321

DROWNED FROM BENEATH— THE HYDROPLATE THEORY 323

THE ANNIHILATION OF LORD LAMECH AND HIS ESTATE .. 326

SUMMATION 329

CHAPTER 14 - THE WORLD THAT PERISHED 335

DINOSAUR EXTINCTION THE RESULT OF A METEOR
IMPACT? 337

WHY AND HOW WERE CONDITIONS DIFFERENT? 339

WHALE-SIZED SAUROPODS 345

BREATHE DEEPLY, LONG-NECK SAUROPODS.........................347

THE SQUARE-CUBE LAW...349

THE CANOPY CONCEPT 352

CHAPTER 15 - WHITTLE AWAY... 363

I, AUTOMOBILE 367

DESIGNED WITH PURPOSE 371

INHERENT PROGRAMING...373

PREPROGRAMED RESPONSE 375

ACCESS GRANTED TO REAMS OF KNOWLEDGE 380

CHAPTER 16 - ORIGINS AND THE TEACHER............................ 389

CHAPTER 17 - HOW LONG WILL YOU WAVER BETWEEN TWO OPINIONS?... 395

WILLFULLY IGNORANT...398

ABOUT THE AUTHOR

Michael Earl Riemer is a poet, skilled machinist, woodcarver, preacher, home Bible study teacher, Sunday school teacher, and author. Born and raised in Milwaukee, Wisconsin, he now makes his home in the nation of the Philippines, on its southernmost island, Mindanao. He resides there with his wife and young son.

He enjoys writing and has authored gospel tracts and articles on various subjects and issues. He has written several books, including *The Path Life Takes*, a collection of poems and short stories, and *Musings on Creation and Evolution*, an assortment of short, well-reasoned arguments refuting evolution. His book *Reindeer Don't Fly* is filled with scientific and logical reasons why the belief in evolution is ill-conceived. *God is One Divine Being* is an engaging study dealing with the Godhead, and his work *ISRAEL, RAPTURE, TRIBULATION: How to Sort Biblical Fact from Theological Fiction* explores the importance of eschatology.

The author's interest in eschatology started shortly before his born-again experience in a Pentecostal church in 1973. It was there that he heard the preaching of an inspiring, "fiery," and well-versed evangelist expounding upon Bible prophecy. The author has been hooked on eschatology ever since.

He is also fascinated by nature and the wonders of creation. For much of his life, he has enjoyed reading about those subjects in books and magazines. However, after reading the Creation story in Genesis and the encapsulated history of the earth presented there, and contrasting it against the backdrop of billions of years of evolutionary history demanded by evolutionists and many other so-called scientists, the author began to examine that subject with greater clarity. With the understanding that evolution is a thinly disguised religion hiding behind the veil of science, he has crafted numerous articles that unmask and expose this dangerous and destructive belief system.

The author is also very concerned about God's Kingdom, our Father's World. He spends some time each week beautifying His Kingdom by picking up trash, pulling out weeds, and planting bushes and flowers. He teaches those who attend his speaking engagements and Sunday school classes to take dominion over each square inch of ground their feet tread upon, for God is interested in everything—each and every activity done under the sun on His planet, throughout each and every culture and nation.

PREFACE

There are many who have faith in evolution who would consider themselves to be intelligent and well-educated. The most likely reason for their belief is the preponderance of propaganda of supposed "facts" and "information" about the primeval earth, its prehistoric creatures, and man's descent, which is found everywhere.

Their indoctrination probably started before they were even in kindergarten, as it did for me. It all began when Mom would read to me my favorite bedtime story, *Dinosaurs*, book number 355, a 25¢ priced edition from the collection of *Little Golden Books* about the long, long ago time before man, when ancient prehistoric creatures called dinosaurs ruled the earth. As a concluding remark on this book's lesson, to drive home its point, its final words clearly state that no living human being has ever seen a live dinosaur.

As I grew older I discovered the immensely popular 48-page *How and Why Wonder Book* series, which many my age will remember fondly. With beautifully illustrated cover art, there were 74 unique titles, including *Prehistoric Mammals*, *Primitive Man*, and probably my all-time favorite children's book, *Dinosaurs*. The storyline was by author Darlene Geis, who, together with illustrator Kenyon Shannon, brought the dinosaurs back to life in a thrilling and exciting way. Priced at 69¢ a book, these volumes were within my parents' budget.

A child's education and instruction in this doctrine, which began on their mother's knee, often continues throughout their formal education and beyond. The students' tutelage is supported by an unlimited number of books, toys, movies, radio reports, and newspapers, all which confirm the validity of this belief. The indoctrination culminates in college, where professors drive home the last nail or two into the lid of the coffin that now holds a student's faith, a once-vibrant devotion to God that had resided within a young child's heart but has now been "educated" out of them. It is pried from their minds bit by bit, "fact" by "fact," until the course of study planned out by the evangelists of Darwinism produces another disciple.

What did Jesus say about the kinds of people who commit an offence to children and purposely lead them astray? *"And whosoever shall offend one of these little ones that believe in me, it is better for him that a millstone were hanged about his neck, and he were cast into the sea"* (Mark 9:42). Jehovah spoke to His people of Israel through the prophet Jeremiah. He told them: *"…and see that it is an evil thing and bitter, that thou hast forsaken the LORD thy God, and that my fear is not in thee, saith the Lord God of hosts"* (Jeremiah 2:19). It is an iniquitous evil, a heinous and bitter thing, this religious doctrine called evolution.

What do you say to your brother or father-in-law when they ask you why you do not believe in evolution? How do you respond to a co-worker who mockingly asks why you believe in fairytales such as those in Book of Genesis, rejecting the "science" of evolution and the evidence of its rationality or validity?

Well, you could make this statement to them and invite their answer: *Could you describe for me briefly the scientific evidence that has convinced you of the validity of the theory of evolution?* No matter if the response comes from your favorite brother-in-law or the tenured science professor, any answer given, even if it consists of "hard facts" that seem to be supplied from "educated" lips, no science, facts or any real evidence will ever be forthcoming.

Is it really true that there are no facts or evidence for evolution? Is it all just a made-up story with no validity? Is evolution a fairy tale like Santa Claus, a yarn or fable that some never learn the truth about, or continue to believe in despite of evidence to the contrary? Well, before the reader finishes this book, just as sure as reindeer don't fly, he will come to the understanding that evolution as taught, contains as much evidence, as a broken cistern contains liquid.

Michael Earl Riemer
San Francisco,
Agusan Del Sur, Philippines
August 9, 2018

Scientists who go about teaching that evolution is a fact of life are great con-men, and the story they are telling may be the greatest hoax ever. In explaining evolution, we do not have one iota of fact.*

INTRODUCTION

Through faith we understand that the worlds were framed by the word of God, so that things which are seen were not made of things which do appear. (Hebrews 11:13)

Where wast thou when I laid the foundations of the earth? Declare, if thou hast understanding. (Job 38:4)

The fear of the LORD is the beginning of knowledge: but fools despise wisdom and instruction. (Proverbs 1:7)

EVERY NEWS STORY POSTED on the Internet or printed in magazines and every book written in our day by evolutionists always starts with the assumption that evolution happened. That belief and assertion is taken as fact and becomes the foundation that the rest of the story is constructed upon.

Evolutionists' *modus operandi*, their primary way of interacting with the public, is analogous to the sleight-of-hand tricks used by all illusionists. It's a bait-and-switch type of ploy that achieves a very powerful effect of which the unsuspecting reader is never aware of when it occurs, nor can identify it when it happens. The ideologues of this religious dogma never divulge to the public when their faith and belief diverge and part ways with facts, truth, and science.

---
*Dr T.N. Tahmisian (Atomic Energy Commission, USA) in 'The Fresno Bee', August 20, 1959. As quoted by N.J. Mitchell, *Evolution and the Emperor's New Clothes*, Roydon publications, UK, 1983, title page

This book explores various aspects of evolution, its history, its dating methods, and teachings. Additionally, at times it will focus directly on the issue of faith: That evolution is NOT science, but its tenets, assumptions, and dogmas are built solely on faith, and that any so-called evidence thereof is always built upon the belief—the "fact"—that evolution is true and nothing else.

Some, such as Ernst Mayr (as we will see shortly) fully make my case. Yet most evolutionists do not seem cognizant of this information, and thus remain blind to the consequence of evolutionary rationale, that it is built totally on belief in a tall tale.

Evolution is a philosophy of life, a religious belief. It is a religion in which faith in its processes is the main tenet. It is a belief from which the scientific methods of testing, replicating, observing, and verifying evidence and proof have been divorced and do not play a part in its study. Back in 2004, an article by Ernst Mayr (a very prominent twentieth-century evolutionary biologist) was published in the November 24th issue of *Scientific American*, titled "Darwin's Influence on Modern Thought." The prefatory statement headlining the article stated:

> Great minds shape the thinking of successive historical periods. Luther and Calvin inspired the Reformation; Locke, Leibniz, Voltaire and Rousseau, the Enlightenment. Modern thought is most dependent on the influence of Charles Darwin.[1]

Mayr proudly extols and exalts the influences, contributions, and major "accomplishments" made by Charles Darwin. The three fields, to which he did not necessarily give birth but generated major intellectual developments and transformations, are evolutionary biology, the philosophy of science, and the modern zeitgeist.

The German word *Zeitgeist* is often attributed to the philosopher Georg Hegel, although he never used the word. In his work *Lectures on the Philosophy of History* he used the phrase *der Geist seiner Zeit*, meaning the spirit of his time. The word encapsulates the idea of the current cultural environment and the conceptual leanings and propensities of the times. Mayr wrote:

> ...no biologist has been responsible for more—and for more drastic—modifications of the average person's worldview than Charles Darwin. [...] Almost every component in modern man's belief system is somehow affected by Darwinian principles.[2]

Mayr also claimed that:

> No educated person any longer questions the validity of the so-called theory of evolution, which we now know to be a simple fact...it has become the basic component of the new philosophy of biology...[3]

According to Mayr, Darwin is responsible for *new* branches of life sciences—evolutionary biology, philosophy of science, and biology. Along with this modern evolutionary conception or notion of biology comes the idea of the common descent of all living species or organisms on earth from a single unique origin; the gradual change of creatures over a very long period of time with no major breaks or discontinuities. The mechanism evolution utilized to accomplish this was natural selection. Mayr continued his line of thought:

> Despite the passing of a century before this new branch of philosophy fully developed, its eventual form is based on Darwinian concepts. For example, Darwin introduced historicity into science. Evolutionary biology, in contrast with physics and chemistry, is a historical science—the evolutionist attempts to explain events and processes that have already taken place. *Laws and experiments are inappropriate techniques for the explication of such events and processes. Instead one constructs a historical narrative, consisting of a tentative reconstruction of the*

> *particular scenario that led to the events one is trying to explain.* (emphasis added)[4]

It is known that Darwin spent many years travelling with nautical expeditions observing a wide variety of wildlife and recording his observations. He was also familiar with works produced by the Royal Society regarding vivisection, and it was only after decades of observation and research that he arrived at his conclusions. Darwin did observe animals in the wild; he studied their niche in nature and current adaptations to their surroundings. However, no amount of study, observation, or surveillance of living animals can determine what their great ancestors may have looked like or how they may have changed over time. You cannot observe the past! Darwin simply believed they did, and then made a wild guess, with its conclusion, that every creature has evolved over long periods of time, starting from simple creatures to more complex. However, his presumption was wrong, for all we have ever observed in nature, even with all the adaptations observed over the years, is that like produces like, the same kind of creature, always, never what Darwin asserted!

Webster's Third New International Dictionary defines historicity as "the quality or state of being historic, esp. as distinct from the mythological or legendary; A condition of being placed in the stream of historical developments." Just how did Darwin accomplish this? How did he turn a biased, prejudiced, and fanciful belief about the past into real history without observation (he never observed the past), repeatable scientific tests, or experiments? How does a belief about the past become historical narrative, fact, or science without unbiased confirmation? Mayr answered:

> The testing of historical narratives implies that the wide gap between science and the humanities...is actually nonexistent—by virtue of its methodology and its *acceptance of the time factor* that makes change possible, evolutionary

biology serves as a bridge. [...] Another aspect of the new philosophy of biology concerns the role of laws. *Laws give way to concepts in Darwinism.* ... These *biological concepts*, and the theories based on them, *cannot be reduced to the laws and theories of the physical sciences.* [...] Observation, comparison and classification, as well as the testing of competing historical narratives, became the methods of evolutionary biology, *outweighing experimentation* ... [Darwin] *established a philosophy of biology by introducing the time factor*, by demonstrating the importance of chance and contingency, and by showing that *theories in evolutionary biology are based on concepts rather than laws*. (emphasis added)[5]

While some may suggest that Mayr was describing how things used to be in the past, and not the current condition of evolutionism, nothing has actually changed and this is still the case.

The word science comes from the Latin Scientia, meaning to know or knowledge. *Webster's New Collegiate Dictionary* defines science as:

> Knowledge attained through study or practice, or knowledge covering general truths of the operation of general laws, esp. as obtained and tested through scientific method [and] concerned with the physical world.

Can evolution be considered a valid scientific theory if it has never been demonstrated that its conclusions are well-founded or defensible by experiments or observation? After reading my question, one evolutionist said this statement was not accurate, and then stated: "There have been many experiments to reproduce the conditions under which life first began on earth ... well-publicized experiments that have successfully produced RNA from "scratch"—Ewen Callaway, "Artificial Molecule Evolves in Lab ..." She continued on and listed a number of other such examples.

This nature of rationalizing and reasoning is an error at the very foundation of evolutionary thinking. This attempted rebuttal clearly demonstrates my contention, that the heart of evolutionism does not beat with science, but pulsates purely with believe and faith in its philosophies. She says many experiments have been carried out to reproduce the conditions under which life first began on earth. That begs the question, how do they know what the conditions were like on earth billions of years ago? Just belief. How do they know for sure, that the Earth, is in fact, billions of years old? Faith in evolutionism. Her answer also assumes that, somehow, life can just start. She also states that experiments have successfully produced RNA from "scratch." Unless I am mistaken, nothing was produced from "scratch," for these experiments were purposely preformed and conducted by intelligent educated men and women, whose education did not happen by chance. They were using labs, equipment, and other paraphernalia that did not just pop into existence, but was created by intelligent and skilled craftsmen. In short, nothing happened by accident, but all was purposely planned and executed.

What is the purpose of scientific study? A generic description would be to construct useful models of reality. There is applied science, which is the application of study and research to human needs. Empirical science is based on or characterized by experiment and sensory observation instead of theory or the application of logic. Operational or observational science increases mankind's knowledge, which allows industry and scientists to produce technology that benefits humanity. The study of evolutionism does not fit within the confines of the definitions of either empirical or operational sciences. Some science educators state the matter this way:

> Recognizing that everyone has presuppositions that shape the way they interpret the evidence is an important step in realizing that historical science is not equal to operational science. Because no one was there to witness the past (except God),

we must interpret it based on a set of starting assumptions. Creationists and evolutionists have the same evidence; they just interpret it within a different framework. Evolution denies the role of God in the universe, and creation accepts His eyewitness account—the Bible—as the foundation for arriving at a correct understanding of the universe.[6]

Mayr makes it very clear that Darwinism rejects all supernatural causations and phenomena. Its tenets eliminate God from science, leaving only "scientific" explanations of all natural phenomena. Darwin's greatest contribution, reasons Mayr, is how the set of new principles influenced the thinking of every person, through evolution; the living world can be explained without recourse to supernaturalism.

Other evolutionists, such as Dr. Scott Todd, echo the same thoughts: "Even if all the data point to an intelligent designer, such an hypothesis is excluded from science because it is not naturalistic."[7] Mayr makes it very clear that Darwinism is based solely on the "*acceptance of the time factor.*" Without this notion of vast ages of deep time, the evolutionists' story is just that, a story, residing in the same realm occupied by genies, flying horses, and magic carpets.

Darwinism deals with biological concepts, not scientific laws. Its basis lies within the sphere of the abstract, the imaginary and speculative empire. Its teachings are arrayed outside the domain and purview of all empirical, operational, or observational science. Mayr reasons that Darwinism is a narrative, a chronicle of life, one in which we must accept as true history. If we could boil down the essence of Darwinism to a few sentences and distil the crux of the matter, these words of Mayr's might be what we are seeking:

> …biologists Thomas Huxley and Ernst Haeckel revealed through rigorous comparative anatomical study that humans and living apes clearly had common ancestry, an assessment that has never again been seriously questioned in science. The

application of the theory of common descent to Man deprived man of his former unique position.[8]

Darwinism rejects all supernatural causalities or wonders, thus depriving man of his rightful and unique position in God's creation. Just as most "-isms" want no competing beliefs or ideas, the religion of Darwinism strives to limit or destroy all others it perceives as rivals or challengers. It hates the truth of Christianity's teaching that Jesus is the Creator of all that is *"All things were made by him; and without him was not any thing made that was made"* (John 1:3), and that He placed man over His creation as its steward.

> What is man, that thou art mindful of Him? and the son of man, that thou visitest him? For thou hast made him a little lower than the angels, and hast crowned him with glory and honour. Thou madest him to have dominion over the works of thy hands; thou hast put all things under his feet. (Psalms 8:4–6)

Maybe at the heart of the matter, because there are only two choices as to where all things came from, *nothing*[9] or a *Creator*, choosing one religion eliminates the other.

NOTES ON INTRODUCTION

1. Ernst Mayr, "Darwin's Influence on Modern Thought," Scientific American, November 24, 2009. https://www.scientificamerican.con/ article/darwins-influence-on modern-thought/
2. Ibid., paras. 1, 32.
3. Ibid., paras. 30, 11.
4. Ibid., para. 5.
5. Ibid., paras. 6, 13, 32.
6. Roger Patterson, *What is Science?* Answers in Genesis web site, last updated July 29, 2014, https://answersingenesis.org/what-is-science/what -is-science/
7. Scott Todd, "A View from Kansas on that Evolution Debate," *Nature* 401(423), September 30, 1999. DOI: 10.1038/46661

8.   Mayr, 2009, para. 25

9.   Gerardus D. Bouw, *Geocentricity: The Biblical Cosmology* (Cleveland, OH: Association for Biblical Astronomy, 1992), p. 321. Bouw provides us with a logical, profound, and weighty reason why there really is only one choice where all things could have come from—*ex nihilo*, which is creation from or out of nothing, something only God can do:

We start at the beginning with absolute nothing. The tendency is to treat "nothing" as a "thing," but its name, "no-thing" belies that. "Nothing" cannot have any properties or attributes. In particular, "nothing" cannot have length, volume, time, or intelligence. It can have neither beginning nor end. It cannot have an origin, and it cannot be a thing. In short, it cannot have the property of existence and so cannot exist. Since it is true, that "absolute nothing" cannot exist anywhere at any time, then its inverse must also be true that "absolute everything" must exist always and everywhere. Now do not confuse absolute everything with absolutely everything. This absolute existence must have all the inverse properties of nothingness. Whereas the nothing has no size, its inverse must be infinite in extent or omnipresent. Whereas nothing has no knowledge, its inverse must be omniscient. Whereas nothing has no existence, its inverse must have infinite existence. Whereas nothing has no power, its inverse is omnipotent. These are precisely the characteristics of God as presented in the scripture. (Note that these characteristics require God to have character or personality also.) Thus we have arrived at the necessity for the existence of God as inferred from the very existence and order built into the universe. This observation also illuminates the error of the Big Bang hypothesis, namely, that the Big-Bang-produced universe is too small and too uncharacteristic to be realistic.

So there was nothing at all before God, and God came from nowhere because there is nowhere God could have come from. Hence God is reasonable and he even invites us to reason with him, for he says: "Come now, let us reason together" in Isaiah 1:18.

It is obvious that radiometric techniques may not be the absolute dating methods that they are claimed to be. Age estimates on a given geological stratum by different radiometric methods are often quite different (sometimes by hundreds of millions of years). There is no absolutely reliable long-term radiological "clock". The uncertainties inherent in radiometric dating are disturbing to geologists and evolutionists...*

CHAPTER 1

HOW OLD IS THE EARTH?

Through faith we understand that the worlds were framed by the word of God, so that things which are seen were not made of things which do appear. (Hebrews 11:13)

WITHOUT A DOUBT, THE subject of ancestries and origins is a topic all rational beings have pondered. Rest assured, the evolutionary philosophy of the origins of the universe and life are teachings that, during a Christian's pilgrimage, will be strewn across his or her path from cradle to grave. Accordingly, our starting point in this first chapter will establish a premise and basis which underpins the arguments the rest of the book is built upon: *the earth is young* (thousands of years old, not billions). That belief and statement rests upon the words recorded in the Book of Exodus 31:17: "*In six days the LORD made heaven and earth, and on the seventh day he rested.*" Therefore, the very beginning

---
*William D. Stansfield, Ph.D. (animal breeding) (Instructor of Biology, California Polytechnic State University) in *The Science of Evolution*, Macmillan, New York, 1977, p 84.

of time took place about 7,000 years ago. That reality leaves some with the question—what was God doing all those billions of years before this recent commencement of history? The simple answer is, those supposed billions of years never transpired; for time did not exist before the creation of heaven and Earth.

This book also rebuffs, rejects, and repudiates the Copernican principle, a.k.a. the mediocrity principle, which was promulgated by Nicolaus Copernicus (1473–1543). This non-proven[1] and non-scientific ideology has had profound effects on the science and philosophy of succeeding centuries. This information is vital if one is to understand the history of modern science, for since the time of Copernicus it has represented a fundamental philosophical change (sometimes called the Copernican Revolution) in how academia dealt with mankind's position and status in the universe and is foundational to all current scientific thought and evolutionary doctrine.

This idea starts with the assumption of mediocrity! This opinion proposes that in the grand scheme of things, our solar system is ordinary, unremarkable, and not unusual in any important way. Nor is our earth favored, or in a central or specially chosen location in the universe. Human beings occupy an ordinary earth and do not have a special, privileged, or exceptional status or position. Carl Sagan in his book *Cosmos* asked:

> Who are we? We find that we live on an insignificant planet of a humdrum star lost in a galaxy tucked away in some forgotten corner of a universe in which there are far more galaxies than people.[2]

In a written correspondence with an evolutionist, she believed I was misrepresenting Sagan. She wrote:

> Sagan's enormous enthusiasm for and wonder at what he saw as the remarkable series of chance occurrences that gave rise to life on earth. In describing life as arising on an ordinary earth

and an ordinary solar system, Sagan and those who agree with his thinking are not denigrating our universe as mediocre, but emphasizing how we should value the fact that it happened here, and learn all we can about it.

My statements in no way imply that Sagan does not have an enthusiasm and awe of our universe and enjoys learning more about it. But unbeknownst to her, she truly makes my point and confirms my contentions and premise. This is not just an ordinary earth, solar system, nor universe that surrounds us (as Sagan and likeminded people believe). Nor did a series of chance occurrences give rise to life. It was not happenstance that brought everything into existence. Our earth is not mediocre, one among tens of thousands which were flung across empty space sometime in the distant past.

Michael Rowan Robinson in his book *Cosmology* emphasized the Copernican principle as the starting point for modern thought: "It is evident that in the post-Copernican era of human history, no well-informed and rational person can imagine that the Earth occupies a unique position in the universe."[3]

Throughout the years this principle has expanded and been adopted into all areas of research, including physics, astronomy, and biology. The argument for extraterrestrial intelligence is also based on the principle of mediocrity. This principle suggests, because the evolution of the solar system and the creation of earth are not unusual in any important way, the same processes that govern the rise of humanity are probably universal, and thus planets capable of spawning life must be common; therefore, in due course lifeforms arose on other Earth-like planets throughout the vast expanse of the cosmos.

Of course they don't know how those processes might work or operate on other planets, or how a difference in gravity might impact life elsewhere. They might debate whether their respiratory functions would be recognizable to us (they might breathe a different gas, they might breathe through different mechanisms, etc.). Some emphasize how rare it might be to find a planet that even has atmospheric conditions that could support life the same way that Earth does. But this whole idea assumes that life can just start (they believe it's a universal fact), and then the process of evolution can take place.

Bouw, in his book *Geocentricity*, relates some very interesting information arising from the effects of this principle:

> One of the interesting side effects of heliocentrism in the seventeenth and eighteenth centuries was the notion that the moon, planets and stars (yes, and even the sun) were inhabited. This idea is reflected in the names of the lunar mare ("seas") and oceans. In fact, the "fact" that the moon was inhabited was at the time considered to be absolute proof against the inspiration and inerrancy of scripture. The reasoning was that with the moon and stars all inhabited, there was nothing special about the earth that God should pay particular attention to events here as opposed to say, the events on the moon or the giant planet Jupiter. Today we know that no place in the solar system has life on it but the earth; but then the belief in the "plurality of worlds" was considered absolute.[4]

It is true that most people no longer believe the moon, sun, and other planets in our solar system are inhabited. However, this idea from centuries past, that life exists elsewhere, someplace in space is still very much alive, and beats strongly within the breast of all evolutionists (and Hollywood movie producers). SETI, the search for extraterrestrial intelligence, the enormous amount of time and money spent on search for life on Mars, and countless other endeavors testifies that the Copernican principle, its attitude and mind-set is still very much alive and has never died.

Contrary to Sagan's statements, the Earth is a special, exceptional, and unique place. It is an oasis in the midst of lifeless planets, moons, and far flung galaxies. It is the exclusive abode, domicile, and residence of all bacterial, animal, and human life. Other than God and His created spiritual beings such as angels, we are alone in the universe. It was only on this earth that the drama of the ages took place. "*God so loved the world, that he gave His only begotten Son, that whosoever believeth in him should not perish, but have everlasting life*" (John 3:16). It is only to us, the sinful beings that live on this Earth, that the Gospel is given: "*Christ died for our sins according to the scriptures: And that he was buried, and that he rose again the third day according to the scriptures*" (I Corinthians 15:3–4).

Carl Sagan's question, "Who are we?" is clearly answered in Scripture:

> What is man, that thou art mindful of Him? and the son of man, that thou visitest him? For thou hast made him a little lower than the angels, and hast crowned him with glory and honour. Thou madest him to have dominion over the works of thy hands; thou hast put all things under his feet. (Psalm 8:4–6)

Before we continue our quest for understanding, for the most part we use the *King James Version* (1611) of the Scriptures. This translation is not a paraphrase, a poetic adaptation such as the *Living Bible*, nor a somewhat loosely translated edition such as the *Good News Bible*. It is a careful rendering of the ancient text. With the exception of the words in italics, which the translators supplied to help the text flow (and so the reader would know they were man's words), each word in the translation has a corresponding counterpart in the original languages. Thus, if you are so inclined to understand the various shades of meaning for the Greek or Hebrew words, Bible study helps such as *Strong's Exhaustive Concordance of the Bible* are available with Hebrew, Chaldee, and Greek dictionaries. This is something that cannot be done with other Bible translations, for in most translations the translators added freely to the Hebrew and Greek texts, thus mingling man's word with God's Word, consequently, the reader will never know if he is reading what God spoke or man's thoughts or ideas (which is the reason that I will, only sparingly, read or refer to other versions). Most translations produced within the last two hundred years or so are unlike the KJV, for the translators did not strive for a word for word translation, but for readability, its thoughts, not fidelity to the ancient text. For example, in the Preface to the 1973 *New International Version* it states:

> The first concern of the translators has been the accuracy of the translation and its *fidelity to the thought of the biblical writers*. They have weighted the significance of the lexical and grammatical details of the Hebrew, Aramaic and Greek texts. At the same time, *they have striven for more than a word-for-word translation*. Because thought patterns and syntax differ from language to language, faithful *communication of the meaning of the writers* of the Bible demands frequent modifications in the sentence structure and constant regard for the contextual meaning of words. (emphasis added)

The ideology and philosophy of present-day scholars in translation of the ancient scrolls; contrasts sharply with those during the days of King James. This is the reason for the numerous "new" versions of Scripture, for in each new translation you will not be reading pure Scripture but commentary, the scholars own interpretation of the text, instead of allowing the text to stand on its own and letting the reader to fathom out what the original author meant. Consider the same portion of Scripture in the original Greek, the KJV, and *Today's English Version* of Revelation 5:5:

> "... behold, overcame the Lion – of the tribe Judah, the root of David..." (Greek text)

> "... behold, the Lion of the tribe of Judah, the Root of David..." (KJV)

"Look! The Lion from Judah's tribe, <u>the great descendant of David</u>." (TEV)

I underlined a portion in these three texts, do you see the difference? When scholars take such liberties with Scripture (as in the TEV), is there any wonder, when two or more Christians read the same verse in different versions, they can come to a completely different understanding, for the same verse in different versions can be literally at odds with each other.

THE BOOK OF GENESIS: LITERAL OR FIGURATIVE HEBREW POETRY?

Let us now delve into the realm of origins and the beginnings of all things. Does the Book of Genesis (or the Scriptures in general) teach or even imply the earth has been around so long, that it has celebrated billions of birthdays? Or does it teach or present our planet as having been around less than 7,000 years give or take a few centuries?

Before you answer either of those questions, please read or have a second look at the Book of Genesis. However, before you do (this will be hard for most people), endeavor to lay aside any bias, and proceed as if you are going to read it for the first time. Grab your Bible, find a quiet spot, and get comfortable. Now go ahead, but focus your attention particularly on the first two chapters with the purpose of understanding the idea the author (Moses, the editor and compiler) was attempting to disclose to the reader. Even if you do not believe it is truth, what is the first impression that comes to your mind? What was Moses attempting to convey to his audience—a long drawn-out creation, one which took an immense amount of time? Was Moses just writing poetry, a "creation hymn," an imaginative limerick that reveals to us in allegorical prose that in the distant unknown past, once upon a time, in an undisclosed fashion, God created everything? When you read these chapters, do you feel Moses is portraying or depicting a non-literal series of seven events, each of unknown duration? Or do you get the distinct sense or perception that a week of normal extent is being communicated to the reader?

Now it may be true, as some have maintained, that during ancient times and eras, writers, historians, and chroniclers had a different sense of historical events and relayed those events a bit differently to their audience. In that case, we must ask what kind of literary style is being employed here. What kind biblical language was used in the first two chapters—figurative Hebrew poetry, or historical chronicle?

The biblical answer is to compare Scripture with Scripture. Moses wrote the account of creation and the words we will now consider. God spoke to Moses, who relayed His directives to the Israelites encamped at the foot of Mount Sinai. "*Six days shalt thou labour, and do all thy work...For in six days the LORD made heaven and earth, the sea, and all that in them is, and rested the seventh day: wherefore the LORD blessed the Sabbath day, and hallowed it*" (Exodus 20:9–11). (see also Exodus 31:17)

As God relates to Moses, man's week is to consist of seven days, just as God employed when He created everything. Out of necessity, these were literal days of 24 hours each; if not, how could long periods of time of unknown duration be used as a pattern for man to divide his time? Just as the last day of the week, the Sabbath Israel was to keep, was a literal day of 24 hours of time, God's rest/ceasing of labor on the seventh day was also literal.

There are a number of views we will focus upon in the next few pages that deny the days in Genesis are literal, and promote the concept of a local flood. For those advocates, the biggest problem is their callous disregard for the plain and straightforward reading of the Scriptures. They question the historicity of the text (only accepting the premise that once upon a time God created everything), and insist that no history of real events and actual occurrences took place, for the first few chapters (some even consider the first eleven chapters) are highly symbolic.

I do understand there is much symbolism used throughout the Scriptures—figures of speech, hyperbole, and so forth, much like the proverbial saying parents bellow at their children: "I told you a million times to stop running in the house!" What many fail to consider is that poetry can be clear as prose and every bit as truthful. Because every word God has spoken in Scripture is truth, the God of Truth can never utter untruths, whether He uses prose or poetry to communicate His message.

Consider the 23rd psalm. It is devotional poetry, filled with biblical metaphors and imagery that illuminate the richness of God's love and care for His people. It speaks truth, clearly and directly to the heart of all who meditate upon its words. Symbolic, emblematic, and illustrative terminology is sprinkled lightly throughout the Holy Scriptures, but it fills the Book of Psalms and writings of the minor and major prophetical books in the Old Testament, as well as in the Book of Revelation in the New Testament.

However, the Book of Genesis through the Book of Ruth belongs in the same category of literature as the other historical Old Testament books such as I and II Samuel, I and II Kings, and I and II Chronicles. These books are not rigid chronology, analytical history, or historical fiction—but neither are they narrative history or traditional narrative literature. When writing historical narrative, authors typically blend facts with imagined characters and situations, similar to what Peter Marshall and David Manuel did in their 1977 book *The Light and the Glory*, a well-crafted historical analysis of the overwhelming influence Scripture had on the founding of America. However, in those Old Testament books, there are no imagined characters, nor are any fictional situations or details interwoven throughout the narratives. You will only read about real people and records of factual events.

From the very first verse in Genesis, the story flows seamlessly, verse after verse, account after account. It is a continuous story, a poignant saga, and a description of real conversations, actions taken, and consequences derived from those behaviors. As surly as Balaam's donkey spoke to him, Elijah raised the son of a widow from the dead, captain

Naaman was cured of leprosy by washing seven times in the Jordan River as stipulated by the prophet Elisha, and God prepared a great fish in which Jonah took up lodging in its belly for three days and nights; the serpent had a two-way conversation with Eve in the Garden of Eden and God asked Cain, "*Where is Abel thy brother?*" (Genesis 4:9).

If Eve's recorded conversation with the serpent is just an imaginary tale, why would the account of God speaking to Cain, and all the other narratives pointed out, be any different? What method would someone use to determine which account is literal and which narratives are fictional? If the first few chapters are not literal, when do real historical reports and events begin to take place? When does the author commence to give us a genuine chronicle and discontinue the figurative and symbolic jargon? If the Book of Genesis, including the first two chapters, is not authentic history, it leaves mankind without a true account of creation and without real answers to life's most basic questions.

How did the earth and the entire universe come into existence (there are of course hypotheses that are based on geology and physics, but as other ideas of this kind, they are ever changing)? Where did life come from? Why do we grow old and die? Why and how did evil come into existence? How did the numerous nations, most with distinct languages, come about or originate? Why did Christ have to shed his blood? Why, where, how, when . . . ? If those verses are not history, there is nowhere else to look for those answers! Furthermore, the only "answers" to those questions one could ever hope to find would be just baseless speculation, such as exhibited in Mike Wehner's article "Scientists Finally Have a Decent Guess as to Why Earth Exists."[5] The Holy Bible is the only book in the entire world that can provide us with reliable and accurate information for these age-old questions.

ADAM, THE FIRST MAN

Was Adam a real person? Do the genealogical records in Scripture start with a phantom? Were Adam's first two sons, Cain and Abel, also apparitions? If so, which one of his immediate descendants listed in Genesis chapter 5 was the *real* thing, a historical person? Adam is listed in Luke's genealogy (Luke 3:38) and in the Book of Chronicles 1:1 as the first man. Jude 14 names him as the first man. He is referred to by name (as the one who caused sin to enter the world) in the Book of Romans 5:14. You will also find him mentioned in I Corinthians 15:22, 45, and in the Apostle Paul's first letter to Timothy, 2:13–14.

Scripture tells us that Adam lived for 930 years (Genesis 5:5). It also lists the names and the age at death for those in the line of descent of Adam, such as Noah, Abraham, Isaac, Joseph, and numerous others. Are the years listed literal years, each comprising 365 days? And are those days which make up literal years comprised of 24 hours each? If not, when do the days and ages listed in Scripture become literal? Where is the demarcation verse, where all the verses before are just figurative, and the verses listed after are literal?

Each of the six creative days had one distinct time of light and darkness, an *evening* and a *morning*. If those creative days were indistinct, long ages, would not everything get quite hot during the long age of daylight, and freeze during the long age of darkness? How would plants survive without light during the long age of night?

Was Adam's first day of life, when he became a living soul, an interminable, unknowable, arcane period of time? Were his second, third, fourth, etc., days more of the same, unknown phases of time? If so, when did Adam's days become literal days of twenty-four hours?

Unless we start from the very first day of Adam's life, counting each day as a normal day of twenty-four hours, it would be impossible to determine the number of years he lived. Because he was created on the sixth day, both the sixth and seventh days would had to have been literal, thus, establishing that each day of the creation week were days of 24 hours each! It is true that the first few chapters in Genesis are not a scientific treatise, nor is the unusual method Jesus used to heal the eyes of a blind man (John 9:6) a recommended medical formula or treatment. Jesus spat on the ground and used the saliva to make mud and anoint the eyes of a blind man. He then sent him to wash in the pool of Siloam, and he received his sight.

In the beginning God created the Earth before the universe. He created light before the sun, moon and stars. He created the chicken before the egg. He created all the fowls of the air and all sea creatures before man and the land animals. Land plants such as fruit trees and seed-bearing plants were made three days before creeping things, which included the pollinating insects such plants require. The seas, teeming with creatures, were here before there was any dry land. Basically, there is no common ground between the way evolution was supposed to occur and the way God ordered the creation.

Throughout this book, I have taken the straightforward, uncompromising view that the Book of Genesis, including its first two chapters, is history. The seven days of creation are not allegorical, symbolic, or poetic devices. Nor does each day represent a long time period of unknown duration or some kind of vague idea or symbolism. Each day was 24 hours and had a distinct period of light and darkness. For those who read the Creation story for the first time, most will come to the same understanding I did, each day is a literal day, and the Creation week is a complete unit of time with no gaps or separation between any of the days.

It was only after I read the Book of Genesis for the first time that I heard about the compromised views we are going to discuss. For most, those views seem badly strained, and a very forced and unnatural reading of Scripture. The other verses the advocates have gathered together for a defense of those doctrines are in extremely short supply, feeble, and generally unconvincing.

The non-biased individual (it may be impossible to be non-biased, for everyone is always influenced by the views of others and information he or she has learned) who reads the story of Creation for the first time would be hard-pressed to point out the verses (or "gaps" between the verses) which some purport and maintain describe long periods of time, countless ages, or millions and billions of years. Such a reader would struggle to find in those first two chapters of Genesis "ape men" who dragged their knuckles through the dirt, had no verbal or written language, and had still not learned the use of fire. Where are the verses located that allow countless eons of time, which are needed for simple organisms to develop or evolve into more sophisticated creatures?

Did you get the impression after reading those chapters that Moses is describing an earth much like ours, where, before Adam came on the scene, mortality, transience, and death had already stalked the land for countless eons? Where are the verses in which the *enemy*, death, caused the demise of an unimaginable amount of vegetation, which allowed it to accumulate and form immense deposits of coal scattered over the face of the earth? In the first two chapters, where is death hiding, something (which at first was not known to Adam and Eve), which caused countless animals to die before Adam sinned? How could God have pronounced everything He made as "*very good*" (Genesis 1:31), when Adam and Eve would have been literally walking on an earth-wide fossil graveyard, a world which contained an immense number of putrefying corpses of animals and man-like creatures which had died from disease, suffering, accidents, and numerous other causes long before God breathed into Adam's nostrils the breath of life and he became a living soul?

You will find none of those things (death, disease, or suffering) in the first two chapters, for they had not occurred at that point. Because those things were not yet in existence, all the fossils of plants and animals (including dinosaurs and other extinct creatures) dispersed throughout the rock layers and coal seams found over the entire earth had to have happened after Adam and Eve were created.

This universe is not billions of years in age, as some Christians contend and all evolutionists insist. Nor do the Scriptures state that the earth is less than 7,000 years in age. Although there has always been some debate by Christians when using Scripture to calculate the age of our earth, the contention dealing with earth's age concerns a few hundred, or at the most a few thousand years of time, not millions or billions of years. While it is not possible to know the precise year and date of creation, there is a means some have employed which will give us a reliable estimate, and fix with relative certainty the age of our earth.

ARCHBISHOP JAMES USSHER

In 1581 James Ussher was born in Dublin, Ireland. While just a young man he committed himself to the work of God. At age 20 he was ordained a deacon and priest in the Anglican Church, and at 26 was appointed chairman of the Department of Divinity at the University of Dublin. From 1607 to 1621 he was a professor and was twice appointed vice-chancellor of Trinity College, Dublin. In 1625 he was appointed to the highest position in the Irish Anglican Church, Archbishop of Armagh. During his lifetime he was appointed to other places of honor, held in great esteem, and was one of the most learned men of his day.

While still a young man he excelled in world and church history, and read every history book he could obtain. As an expert in Semitic languages, he argued for the reliability of the Hebrew text of the Old Testament. He wrote extensively in both Latin and English on Bible-related topics and had a magnificent library of 10,000 volumes, which is now housed in the University of Dublin, Ireland.

In 1650 Archbishop Ussher completed a 1,600-page document in Latin, a complete history of the world which covered every major event from the time of creation to A.D. 70. In the 1650s he put forth his scholarly thesis that Creation began on October 23, 4004 BC. This year appeared as a marginal note in most Bibles until the end of the twentieth century.

Ussher began his studies with the assumption that God's Word is true and reliable, thus he believed that the Bible was the only reliable source document of chronological information for the time periods covered in Scripture. For his calculations he mainly used the Book of Genesis, particularly the chronologies listed in chapters 5 and 11. In the original document, his detailed calculations consumed over 100 pages.

In the 2003 edition of Ussher's *The Annals of the World*, revised and updated by Larry and Marion Pierce, Appendix B relates some of the reasons and explanations of how the date of Creation was determined. In the Editor's Preface of this revised book they state:

> In most history books, it is very difficult to tell where the material came from ... This is not true of Ussher's work. It contains more than twelve thousand footnotes from secular sources and over two thousand quotes from the Bible or the Apocrypha. There is very little editorializing and most editorial comments come from the original writers themselves. We were able to verify about 85 percent of the footnotes pertaining to secular history.[6]

Though some may not agree with the date by the late Archbishop, this information certainly shows the great detail and time he spent doing research on this issue. He was a scholar of immense learning, had uncompromising faith in Scripture, and was a man whose research, opinions, beliefs, and conclusions cannot be easily laid aside.

Henry M. Morris, in his superb 1976 commentary on the Book of Genesis, *The Genesis Record*, relays a list of others who also attempted to date creation: Jewish, 3760; Septuagint, 5270; Josephus, 55555; Kepler, 3993; Melanchthon, 3964; Luther, 3961; Lightfoot, 3960; Hales, 5402; Playfair, 4008; and Lipman, 3916. All the dates are different, but most are still reasonably close to Ussher's.

In 1961, John C. Whitcomb, Jr. and Henry M. Morris published *The Genesis Flood*. In Appendix II, the authors devote fifteen pages exploring and discussing the problems, difficulties, and pitfalls associated with constructing a biblical chronology. In their conclusion they state:

> Some evangelical scholars, seeing the possibility of gaps in the genealogy of Genesis 11, have urged an acceptance of uniformitarian and evolutionary dating schemes for early man ... Evangelical scholars who feel the necessity of bringing Genesis 11 into conformity with current paleoanthropological

timetables should realize the full implications of such harmonization efforts... The time has come when those who take the testimony of God's infallible Word with seriousness should begin to look with favor upon the efforts of those who are examining and exposing the unwarranted assumption and false presuppositions of uniformitarianism as it applies to the dating of early man.[7]

It would be a formidable and time-consuming task to date creation using just the Scriptures. Seeing I am not as learned, nor as inclined to do so (but because I understand the method), I agree with these noble men, that about 6,000 to 7,000 years ago God created the Heaven and the Earth.

Ussher's chronology and scholarship have been contested by some. His accepted chronology was based upon an assumption there are no gaps in the biblical genealogies of Genesis 5 and 11. William Henry Green, an Old Testament professor at Princeton Theological Seminary tackled this assumption in his 1890 article in the journal *Bibliotheca Sacra*, "Primeval Chronology." After Green examined the biblical genealogies he found that "there is an element of uncertainty in a computation of time which rests upon genealogies, as the sacred chronology so largely does. Who is to certify us that the antediluvian and ante-Abrahamic genealogies have not been condensed in the same manner as the post-Abrahamic?" He closed his thorough article with these words: "On these various grounds we conclude that the Scriptures furnish no data for a chronological computation prior to the life of Abraham; and that the Mosaic records do not fix and were not intended to fix the precise date either of the Flood or of the creation of the world."

Even if there are a few gaps within the record, and Professor Green was correct, tens of thousands or millions of years cannot be inserted within those gaps and still leave the Scriptures with any semblance of accuracy or history.

So why was it abandoned and discarded? It fell into disrepute, not over the time-honored belief in literal days, but the acceptance of man's word, the findings of evolutionary supporters, instead of the clear teachings of Scripture. The cause and origin for dismissal of Ussher's chronology is unbelief.

> By the middle of the 19th century, Ussher's chronology came under increasing attack from supporters of uniformitarianism, who argued that Ussher's "young Earth" was incompatible with the increasingly accepted view of an Earth much more ancient than Ussher's. It became generally accepted that the Earth was tens, perhaps even hundreds of millions of years old. Ussher fell into disrepute among theologians as well; in 1890, Princeton professor William Henry Green wrote a highly influential article in *Bibliotheca Sacra* entitled "Primeval Chronology" in which he strongly criticised Ussher. He concluded:
> "We conclude that the Scriptures furnish no data for a chronological computation prior to the life of Abraham; and that the Mosaic records do not fix and were not intended to fix the precise date either of the Flood or of the creation of the world."[8]

As we will show throughout this book, there is no evidence that evolution happened or could happen. We will show beyond a reasonable doubt that *belief* is the only "proof" the earth is billions of years old, for there is *no* reliable scientific method to determine earth's age. We will continue to explore the question of why any Christian would believe the earth is billions of years in age, when this assertion is not backed by science, nor clearly taught in Scripture. The answer appears to be as stated: Instead of using Scripture as the starting point, some have let man's ideas creep into their minds and attempt to make Scripture fit their "science" or geology.

Remember, any old earth proofs or studies always start with the assumption that the earth and universe are billions of years old, and those "old earth" believing Christians also have to assume beforehand that this is the case. They then go to the Scriptures and attempt to reinterpret them to fit the evolutionary model, which model would never stand without long ages of time. In effect, they give evolutionists a foothold, a purchase upon which to start their tale. Whereas, if someone is a Christian, it truly makes no difference whether they are studying science, mathematics, or geology, the Bible is authoritative in all matters that it touches upon: "*All scripture is given by inspiration of God, and is profitable for doctrine, for reproof, for correction, for instruction in righteousness*" (II Timothy 3:16).

DAY-AGE THEORY

There have been a number of attempts to harmonize Genesis with evolutionary geology. We will deal with the two most popular, the Day-Age Theory and the Gap Theory, also known as the Ruin-Reconstruction Theory. For those who hold to either evolutionary compromise, each "theory" has a number of variations, interpretations, and nuances. We will start with the Day-Age Theory.

For a much more comprehensive look and discussion of both these schemes, for your review you can find these articles online.

- Dr. Robert V. McCabe, Professor of Old Testament, Detroit Baptist Theological Seminary, "Old Testament Studies: What About the Gap Theory?"

- Bible Answer Man Hank Hanegraaff, "The Gap Theory of Genesis 1:2."

- Ken Ham, The New Answers Book 1, Chapter 5, "What About the Gap & Ruin-Reconstruction Theories?" September 6, 2007.

- "What's Wrong with Believing in the Day-Age Theory?" Creation Moments web site, April 19, 2017,

- Henry M. Morris, Ph.D., "The Day-Age Theory Revisited," Institute for Creation Research website.

The basic Day-Age Theory teaches that each creation day lasted millions of years, and "overlapped" the others by various amounts. Each day was a long creative process in which God at times would "pop" some creature into existence or use evolution to help move things along.

Evolutionary geologists insist that long ages of time were needed to lay down all the strata, such as we find in the Grand Canyon, so each creation day became millions of years in length to provide this time. Thus, they reject the flood of Noah as the sole cause of those layers; indeed, most deny that a global flood occurred, insisting that it was just a local or regional inundation. There are numerous scientific and Scriptural problems with this idea. We have already touched upon one problem with the "science"—the order of creation. Let us then turn to the Scriptural issues. First of all, in the entire Scripture, whenever the word day is modified by the use of the numerical adjective (first day, second day, etc.), it always means a period of 24 hours. While there is an abundance of other problems with this view, we will focus in on just one other conundrum. If the flood was local, why did God instruct Noah to spend what was probably years building an ark? Surely in just a few weeks or maybe a few months he and his family could have migrated well beyond the deluge's ground zero and its surrounding enclaves (most local-Flood advocates maintain the flood occurred in the Middle East, possibly the Mesopotamian region).

Some authors who subscribe to this view, such as Timothy P. Martin and Jeffrey L. Vaughn, object to my answer. They stated:

> Rather than presenting a problem for a local flood view, this question exposed how young-earth creationism's plain literal priority in reading the account entirely misses the biblical emphasis of the account. God planned the events to picture judgment and salvation. There is a spiritual dimension to the story, because the ark is a picture of Christ. What young-earth creationists often miss in their zeal to defend a plain-literal reading of the story of Noah's ark is that it is not about the

> geological history of the planet Earth. It is about the gospel of Jesus Christ. This is made plain by how Peter used the flood event in 1 Peter 3:21–22.
>
> In God's plan it was important that Noah enter the ark as an "incarnation" of the gospel; Noah rested in Jesus Christ for salvation. Noah was figuratively "in Christ" while he was "in the ark." God has a plan whenever he gives his servant a mission, whether it is Noah, Abraham, Ezekiel, or Hosea. Any speculation that wanders from the redemptive purposes of God has lost touch with the biblical emphasis.[9]

Have those who teach a world-wide flood, as plainly conveyed in Scripture, lost touch with and missed the biblical emphasis of the account? Is not the message recorded: Build an ark, for a judgment in the form of a flood is coming, and all those who are not on board the boat will drown? Also, make sure to take a representative pair of all land animals, for all the other beasts are going to drown!

> I do bring a flood of waters upon the earth, to destroy all flesh, wherein is the breath of life, from under heaven; and every thing that is in the earth shall die ... And of every living thing of all flesh, two of every sort shalt thou bring into the ark, to keep them alive with thee; they shall be male and female. (Genesis 6:17–19)

Of course you can apply a spiritual dimension to the story. Paul in writing to the Church in Corinth used Israel's 40 years of travel through the wilderness and their times of disobedience as spiritual lessons, as types and examples for our admonition:

> Neither be ye idolaters, as some of them ... Neither let us commit fornication, as some of them committed ... Neither let us tempt Christ, as some of them also tempted ... Neither murmur ye, as some of them also murmured, and were

destroyed of the destroyer. Now all these things happened unto them for ensamples: and they are written for our admonition. (I Corinthians 10:7–11)

Peter in his epistle also uses a similar typology to communicate his teaching. Like Paul, his analogy is taken from actual events, things which took place. He even said the flood was a *figure* of judgment and salvation through Christ: "*The like figure whereunto even baptism doth also now save us (not the putting away of the filth of the flesh, but the answer of a good conscience toward God,) by the resurrection of Jesus Christ*" (I Peter 3:21).

They are right; the account is not about the geological history of planet Earth. The flood is a story of God's judgment upon all of rebellious mankind. The purpose of the flood was to destroy (by drowning) every single living human being (and all the animals). "*And God said unto Noah, The end of all flesh is come before me; for the earth is filled with violence through them; and behold, I will destroy them with the earth*" (Genesis 6:13).

Some still balk at and reject a universal flood which destroyed all of mankind. Martin and Vaughn stated:

> Here is a real problem for a global-flood view. Eight people on the ark are not enough to maintain these technologies along with the technologies that supported them and carry them all into the post-flood world. Eight people are not enough to develop these technologies anew after the flood. With only eight people, these specialties would have been lost at the time of the flood. They would then have needed to be re-invented when the population recovered… The author of Genesis works from the assumption that musical instruments, metallurgy, etc., *existed continuously from before the flood to his present day.* The economics of the situation implies that at least some of Cain's line lived beyond the confines of the event. God judged a covenant world, the line of Seth, for their

apostasy. This explains why there is no break in previously developed civilized technology at the time of the flood event. Basic economics points to a local flood.[10]

I will deal with a few of their contentions concerning the loss of technologies after the flood in the coming chapters. They reason that because of economics, some made it out alive. The Scripture does not imply a few survived the flood, but emphatically, in clear and precise language, describes a devastating deluge in which all not aboard the ark perished. Their argument that the flood was local and came upon a single group of people, the descendants of Seth for their apostasy, is extremely hard to grasp in the light of Scripture. After reading the following verses, can the reader find any hint, insinuation, suggestion, or mention of any such idea?

> [5]And God saw that the wickedness of man was great in the earth, and that every imagination of the thoughts of his heart was only evil continually. [6]And it repented the LORD that he had made man on the earth, and it grieved him at his heart. [7]And the LORD said, I will destroy man whom I have created from the face of the earth; both man, and beast, and the creeping thing, and the fowls of the air; for it repenteth me that I have made them. (Genesis 6:5–7)
>
> [11]The earth also was corrupt before God, and the earth was filled with violence. [12]And God looked upon the earth, and, behold, it was corrupt; for all flesh had corrupted his way upon the earth. [13]And God said unto Noah, The end of all flesh is come before me; for the earth is filled with violence through them; and behold I will destroy them with the earth. (Genesis 6:11–13)
>
> [17]And, behold, I, even I, do bring a flood of waters upon the earth, to destroy all flesh, wherein is the breath of life, from under heaven; and every thing that is in the earth shall die. (Genesis 6:17)

> ⁴I will cause it to rain upon the earth forty days and forty nights; and every living substance that I have made will I destroy from off the face of the earth. (Genesis 7:4)
>
> ²¹And all flesh died that moved upon the earth, both of fowl, and of cattle, and of beast, and of every creeping thing that creepeth upon the earth, and every man; ²²All in whose nostrils was the breath of life, of all that was in the dry land, died. ²³And every living substance was destroyed which was upon the face of the ground, both man, and cattle, and the creeping things, and the fowl of the heaven; and they were destroyed from the earth: and Noah only remained alive, and they that were with him in the ark. (Genesis 7:21–23)

Clearly, without controversy, these Scriptures present the flood as global, a total inundation, which covered all the hills and mountains and killed not just the bloodline of Seth, but everyone except those on the ark. If it was just a local flood, there would have had to be some kind of valley to contain the water, for water seeks its own level, and does so quickly. Out of necessity, water, by its own nature, if not confined, would have flowed over the entire earth and covered all the other mountains.

So, these local flood advocates expect us to believe that after 1,656 years, all of mankind (even the bloodline of Seth) was still enclosed within a limited locality, a large valley, river basin, or deep depression upon the landscape? Unless this valley was large enough to encompass the entire earth, this idea is ridiculous. Another aspect that must be considered is the question of how many people perished in the flood. If there was a relatively small population, it might be reasonable to assume that mankind (including Seth's descendants) had not spread out much beyond their original ancestral home. However, given the nature of mankind at the beginning, all physical capacities were still running at nearly 100 percent, including their reproductive systems. Regarding many of the patriarchs, including Adam, Seth, Methuselah, and others the Scriptures state: "*he begat sons and daughters*" (Genesis 5: 4, 7, 10,

13, 16, etc.). Therefore, using an extremely conservative population increase of 1.5 percent per year for 1,656 years, there would have been 774,000,000 people alive at the time of the flood!

All mankind is endowed with curiosity and an innate urge to explore their surroundings. Anyone who has spent just a bit of time around toddlers knows their propensity to explore and "get into things." As they grow into teenagers and beyond, this urge and inclination to see the sights still resides within. The Age of Exploration or the Age of Discovery started around the end of the fifteenth century and lasted until the eighteenth century. However, the compulsion to explore began long before that time, for there is good evidence of global exploration from earlier periods. Centuries before Columbus sailed the ocean blue, the Vikings, Romans, and maybe even the Chinese, among others, had set foot upon the shores of North and South America.[11]

Even long after Adam's sin, the curse upon the ground and its aftereffects, this earth, with its forests, landscapes, and wonderful terrain, would have been a place of astonishing beauty and wonder before the flood. Its allure would have been irresistible to Adam's descendants, and mankind would have ventured out into, surveyed, and explored every part of it. They would have colonized and taken possession of every corner of this planet, and constructed a civilization with cities and towns. Therefore the flood must have been a world-wide event and covered the entire earth, or some of the population living in the outlying territories or distant lands would have been left untouched by flood waters (if it was a local flood), which filled up just a single large valley or river basin located tens of thousands of miles away.

> And the waters prevailed, and were increased greatly upon the earth; and the ark went upon the face of the waters. And the waters prevailed exceedingly upon the earth; and all the high hills, that were under the whole heaven, were covered. Fifteen cubits upward did the waters prevail; and the mountains were covered. (Genesis 7:18–20)

There are a few other things about a local flood that makes no sense. Tell me, have you ever heard of a year-long local flood? The waters of the flood covered all the high hills and mountains for about seven months. Noah and his family stayed aboard the ark for over a year before the waters receded sufficiently, and allowed the earth to dry before they could disembark.

What would a local flood have been like? Picture a large bowl. This bowl represents the valley which was occupied by the descendants of Adam. This valley was filled by the waters of the deluge. When those flood waters get to the top of the valley (rim), they will overflow and run down the sides. The circumference of that rim must have been hundreds or thousands of miles in length. The flood waters covered all the mountains (which would have necessarily included the total length of the rim), over 20 ft deep for months. How did the flood waters maintain their depth on top of the rim for months (even after the rain stopped), when they would have poured off the rim within just a few seconds?

If this flood took place in a river valley, what kept the water from running out as fast as it came in? Rivers certainly overflow their banks, sometimes by a massive amount, but as soon as the rain stops, the water immediately starts to drain. What is most peculiar for this local flood, after the 40 days of rain stopped, how did the waters in this valley keep rising for months, without draining? Perhaps this was accomplished by using two mammoth-sized corks, one placed at each end of the valley?

Not only was the flood sent to destroy all of mankind, but all the land animals as well.

> And the LORD said, I will destroy man whom I have created form the face of the earth; both man, and beast, and the creeping thing, and the fowls of the air; for it repenteth me that I have made them. (Genesis 6:7)

Wild rabbits, an invasive pest introduced to Australia in the late eighteenth century, now inhabit most of that continent. Even the rabbit-proof fence that was completed in 1907 did little to stem their advance. So even if mankind had not colonized the entire earth, a global flood was needed, for the animals most certainly would have migrated into places not yet occupied.

RUIN-RECONSTRUCTION THEORY (GAP THEORY)

Let us now reflect upon one other evolutionary compromise, the gap theory.

Billions of years ago, God created a perfect earth with animals and a race of man-like beings that had no souls. Nothing is known about the process of this creation or how long it took. This unspoiled earth existed for an unknown period of time. Satan, the ruler of the earth, dwelled in a Garden-of-Eden-type place (Ezek. 28) and desired to be like God (Isa. 14), so he rebelled. Through his rebellion, sin entered the universe, which brought upon the earth God's judgment in the form of a flood (Lucifer's Flood). After this judgment some even believe there was a global ice age, for the sun lost its power to heat the earth. Any fossil remains of plant, animal or human found today date from "Lucifer's flood." Therefore there is no genetic relationship between those fossils and any living creature upon the earth today.

This modern doctrine was first introduced in Europe by the Scottish theologian Thomas Chalmers, who proposed it in a lecture in 1814. Chalmers' view was picked up and popularized by Plymouth Brethren writer G. H. Pember in his 1876 book *Earth's Earliest Ages*. (Numerous editions of this work were published; the fifteenth came out in 1942). The view had great success in America when it was adopted by C. I. Scofield and included in the notes of the *Scofield Reference Bible* (first published in 1906) and other Bible aids such as *Dake's Annotated Reference Bible* and *The Newberry Reference Bible*. It still remains entrenched as a popular orthodoxy among many dispensational Christians.

All advocates of the gap theory agree with these fundamentals:

- Genesis 1:1 describes a complete and perfect creation of the heavens and the earth.
- Genesis 1:2 records the ruin of the original perfect earth, and an elapsed time period of unknown duration (perhaps billions of years of geologic time) between the perfect earth and its restoration and recreation as set forth in Genesis 1:3–31.
- Harmonization of a literal view of Genesis with a long but unidentified age for the earth.
- A compulsion to place the origin of most of the geologic strata and other geologic evidence between Genesis 1:1 and 1:2.

All of these tenets and stipulations supposedly pour forth from just a few words (bold font) in Genesis 1:2. "*And the earth was **without form, and void** and darkness was upon the face of the deep. And the spirit of God moved upon the face of the waters.*" This Hebrew phrase, tōhû vabōhû, is rendered in most English Bibles as "without form and void." There are a few more verses which were not cited, that those who take hold of this doctrine marshal in defense and argument. There is much more that can be written in explanation and discussion of this idea. But I believe enough has been relayed so as to give the reader a basic understanding of both teachings. There are other teachings in a similar vein, such as progressive creation and theistic evolution, but all have one thing in common: they attempt to harmonize the Scripture with evolutionism's teaching of uniformitarianism.

Uniformitarianism is a concept which claims the present is the key to the past. It maintains that earth's current surface features are a result of slow-moving processes of nature, and that these processes were the same in the past as what is currently observed.

AN UNNECESSARY COMPROMISE

In the remaining portion of this chapter we will continue to explore the reason why, during the last 200 years or so, some Christians have felt the need to compromise the Scriptures. What caused them to abandon the clear and plainly written account of Creation in the Book of Genesis and adhere to evolutionism's belief in long ages of time?

It is always enlightening when one understands the origins, roots, and reasons a particular doctrine was birthed. This is certainly true of evolutionary geology. Read what Dorsey Hager stated:

> The most important responsibilities of the geologists involve the effect of their findings on the mental and spiritual lives of mankind. Early geologists fought to free people from the myths of Biblical creation. Many millions still live in mental bondage controlled by ignorant ranters who accept the Bible as the last word in science, and accept Archbishop Ussher's claim that the earth was created 4004 B.C. Attempts to reconcile Genesis with geology lead to numerous contradictions. Also the theory of evolution greatly affects modern thinking. Man's rise from simple life forms even today causes much controversy among "fundamentalists" who cling to a literal belief in the Bible.[12]

It is somewhat disconcerting that those who dreamt up long ages of time for the express propose of mythologizing the account of creation in the Book of Genesis now find they have allies and staunch champions (double agents?) within the Christian camp. Evolutionism is a malevolent occupying force, whose armies have captured the minds, ideology, and dogma of much of academia. Therefore, those who maintain they are Christian, but at the same time hold to, affirm, and unreservedly propagate its essential doctrine and tenet, millions and billions of years; would they not be considered collaborators and quislings for siding with those who hate truth? A bit of Scriptural discernment and admonition is in order.

> Ye cannot drink the cup of the Lord, and the cup of devils: ye cannot be partakers of the Lord's table, and of the table of devils. (I Corinthians 10: 21)

> For if the trumpet give an uncertain sound, who shall prepare himself to the battle. (I Corinthians 14:8)

> And have no fellowship with the unfruitful works of darkness, but rather reprove them. (Ephesians 5:11)

In Wikipedia there is a discussion of English theologian George Hawkins Pember's 1876 book, *Earth's Earliest Ages*. It states:

> In this book Pember attempted to reconcile the Genesis account of the world's creation with the emerging fossil evidence in geological science about the age of the earth. Pember argued a position known as "The Gap Theory"... In this theory, God originally created the universe but due to the rebellion of some angels led by Lucifer (or Satan) the earth descended into chaos and life was destroyed... So proponents of the Gap Theory like Pember propose that a "gap" exists between the first two verses in Genesis chapter one which allows for all the extra time needed to include the ancient fossil and geological evidences. The geological fossils were creatures that lived in the original creation and were destroyed when Lucifer fell into sin. The biblical story of Adam and Eve is about a later recreation of the world. Pember's argument for the "Gap Theory" is an example of how some evangelical Christians in the nineteenth century tried to reconcile geological evidence for an old earth with the book of Genesis and without embracing Charles Darwin's theory about the evolution of the species.[13]

It seems Pember accepted what those early unbelieving geologists alleged concerning the age of the earth, thus he felt the need to combine this new "science" with Scripture.

The *Creation Moments* website asks the question: What's wrong with believing in the Day-Age Theory?

Somehow, to believe that the Earth was created millions of years in the past seems more rational and is perceived to be supported by the findings of science. The Scottish popular writer Hugh Miller [1802–1856] was an evangelical Christian and very familiar with rocks and fossils. In his book, *The Testimony of the Rocks*, he argued that the creation days spoken of in Genesis chapter one were actually the days when Moses received his revelation of the creation on Mount Sinai! In this way, Miller believed he had reconciled the few thousand years implied by Scripture with the millions of years demanded by geology. After completing his manuscript, he was deeply troubled and shot himself on Christmas Eve, 1856. His book appeared from the publisher the following month. His was one of the first Day-Age theories and was short-lived.[14]

On the fine website *Apologetics Press*, there is a post by Bert Thompson, Ph.D., "Popular Compromises of Creation—The Day-Age Theory." In his discussion he stated:

> John Klotz addressed this point in *Genes, Genesis, and Evolution*: "It is hardly conceivable that anyone would question the interpretation of these as ordinary days were it not for the fact that people are attempting to reconcile Genesis and evolution" (1955, p. 87). [...]
>
> [Professors Donald England and Jack Wood Sears] explained in detail in their writings [that] uniformitarian dating methods take precedence over the Bible! Scientific theory has become the father of biblical exegesis. The question being asked is not, "What does the Bible say?," but rather, "What do evolutionary dating methods indicate?" In order to force the biblical record to accommodate geologic time, defenders of these dating methods do indeed find it necessary to invent "long, complicated, and imaginative" theories.[15]

We will consider one more article, "What About the Gap Theory?" by Dr. Robert V. McCabe, Professor of Old Testament for the Detroit Baptist Theological Seminary:

> Before the development of geology in the late eighteenth and early nineteenth centuries, Christians had explained that the earth's sedimentary rocks containing fossils of once-living creatures were results of the Noahic Flood... However, with the rise of scientific geology, the sedimentary strata and fossil remains received a new uniformitarian explanation... Recognizing the challenge that uniformitarianism presented to orthodox Christianity, Thomas Chalmers of Scotland sought to harmonize Scripture and science...[16]

Professor McCabe continues his discussion with an 1814 lecture by Chalmers on the gap theory and concludes with:

> The gap theory had a great appeal for earlier fundamentalists. This position had at least two strengths. First, the gap theory allowed Bible-believing Christians to affirm what was patently obvious in Genesis 1; viz., the creation account of Genesis 1 took place on six literal days. Second, one could harmonize a literal interpretation of Genesis 1 and the rest of the Bible with the apparently indisputable facts of geological research.[17]

Ever since the idea that the fossil record encompasses millions of years became popular in the early nineteenth century, it has been a struggle for some Christian leaders, and many still wrestle with it. However, the way some have attempted to harmonize evolutionism's geology with Scripture by inventing and concocting doctrines that compromise its integrity and reliability has been detrimental and damaging to the Church.

What is significantly bizarre and farcical about the whole idea and notion is that there is absolutely no need to do so! For in reality, they are trying to harmonize something which never existed, a figment of someone's imagination (billions of years) with truth. As we will show throughout this book, evolution cannot occur, and there is no scientific method that can measure the age of the earth! Evolutionism's speculation of billions of birthdays for our Earth is built on just one thing, and it isn't science—it is *faith*.

Just as sure as reindeer don't fly, billions of years of time never happened. The only place they have ever existed, and ever will exist, is in the mind of an evolutionist.

NOTES ON CHAPTER 1

1. While this book will not cover the arguments for and against the debate between heliocentrism versus geocentricity, we will dispel one myth. Erroneously, many have been taught that Copernicus made discoveries which lead to corrections in astronomy that proved the heliocentric view of the universe. This is simply not true. Numerous statements from physicists concur with this quote by Sir Fred Hoyle: "We know that the difference between a heliocentric theory and a geocentric theory is one of relative motion only, and that such a difference has no physical significance." As Bouw points out in his book *Geocentricity*: "If anyone should know whether or not science has proven heliocentrism, it would be someone of Hoyle's stature" (Bouw, 1992, p. 330).

    Fred Hoyle also stated: "The relation of the two pictures [geocentricity and heliocentricity] is reduced to a mere coordinate transformation and it is the main tenet of the Einstein theory that any two ways of looking at the world which are related to each other by a coordinate transformation are entirely equivalent from a physical point of view....Today we cannot say that the Copernican theory is 'right' and the Ptolemaic theory 'wrong' in any meaningful physical sense." *Nicolaus Copernicus* (London: Heinemann Educational Books Ltd., 1973), p. 78, http://www.icr.org/article/geocentricity-creation/ Fred Hoyle.
2. Carl Sagan, *Cosmos* (New York, NY: Random House, 1980), p. 193.
3. Michael Rowan-Robinson, *Cosmology* (3rd ed.) (London: Oxford University Press, 1996), pp. 62–63.
4. Bouw, 1992, p. 202.
5. In his May 14, 2017 column, "Scientists Finally Have a Decent Guess as to Why Earth Exists," Wehner stated: "Earth is a pretty nifty place...but why is it here at all? For a long time, researchers have tried to answer that question with varying degrees of success, but a new theory of how Earth formed is gaining traction, and it might be the explanation we've been looking for."

To answer this query, Dr. Alexander Hubbard, Ph.D., who works with the American Museum of Natural History, has proposed a new answer. In the beginning:

> ...the sun went through a period of intense volatility in which essentially roasted much of the material in its immediate vicinity, stretching as far as Mars. The softened materials would have been the right consistency to bunch up and form planets, and would explain why the rocky worlds of Mercury, Venus, Earth and Mars sprung up. (Mike Wehner, "Scientists Finally Have a Decent Guess as to Why Earth Exists," BGR News, May 13, 2017, https://finance.yahoo.com/news/scientists-finally-decent-guesswhy-earth-exists-160014149.html, para. 3)

As with all such proposals, it is only good for a limited time, until a new and improved concept takes its place.

6. James Ussher, *The Annals of the World*, edited by Larry Pierce and Marion Pierce (Green Forest, AR: Master Books, 2003), p. 5.
7. John Whitcomb and Henry Morris, *The Genesis Flood* (Philipsburg, NJ: Presbyterian and Reformed Publishing, 1961), p. 489.
8. "Ussher Chronology," *Wikipedia*, July 3, 2017, https://en.wikipedia.org/wiki/Ussher chronology, paras. 16–17.
9. Timothy Martin and Jeffrey Vaughn, *Beyond Creation Science: New Covenant Creation from Genesis to Revelation* (Whitehall, MT: Apocalyptic Vision Press, 2007), p. 458.
10. Ibid, p. 121.
11. Jill Withrow Baker, "The Maps that Columbus Used," n.d., http://www.academia.edu/8586182/The_Maps_that_Columbus_Used.
12. Dorsey Hager, "Fifty Years of Progress in Geology," *Geotimes*, 2(2), August 1957, p. 12.
13. "G. H. Pember," *Wikipedia*, June 21, 2017, https://en.wikipedia.org/wiki/G._H._Pember, para. 5.
14. "What's Wrong with Believing in the Day-Age Theory?" Creation Moments web site, n.d., http://www.creationmoments.com/content/whats-wrong-believing-day-age-theory, para. 1.
15. Bert Thompson, "Popular Compromises of Creation—The Day-Age Theory," Apologetics Press, 1994, https://www.apologeticspress.org/apcontent.aspx?category=9&article=391, paras. 9, 13.
16. Robert McCabe, "What about the Gap Theory?" Old Testament Studies website, n.d., http://www.oldtestamentstudies.org/my-papers/other-papers/recent creationism/what-about-the-gap-theory/, para. 1.
17. Ibid.

It is therefore a matter of faith on the part of the biologist that biogenesis did occur and he can choose whatever method of biogenesis happens to suit him personally; the evidence for what did happen is not available.*

CHAPTER 2
FROG PUREE—THE STUFF OF LIFE

But ask now the beasts, and they shall teach thee; and the fowls of the air, and they shall tell thee: Or speak to the earth, and it shall teach thee: and the fishes of the sea shall declare unto thee. Who knoweth not in all these that the hand of the LORD hath wrought this? In whose hand is the soul of every living thing, and the breath of all mankind. (Job 12:7–10)

PROPAGANDA IN THE TEXTBOOKS

EVOLUTIONISTS WANT THE REST of us, the "uninformed," to believe as they do: That evolution is the "cornerstone" of modern biology. A few years ago the National Academy of Sciences (NAS) published a guidebook entitled *Teaching about Evolution and the Nature of Science*, and made it available to educators throughout America. Its Preface states:

> Many students receive little or no exposure to the most important concept in modern biology, a concept essential

---
*Prof. G. A. Kerkut (Department of Physiology and Bio-chemistry, University of Southampton) in *Implications of Evolution*, Pergamon Press, London, 1960, p. 150.

to understanding key aspects of living things—biological evolution.¹

Their statement that many students receive little or no exposure to the concept of biological evolution is propaganda. As evidenced by my Preface, as air envelops the body, *all* media—and the whole secular education system in America, the Philippines, and most other countries around the earth—is buttressed by evolution. Other than Christian books, I have never seen a science text book, when dealing with this subject, which does not present this idea in words similar as are found in this science book used in high school in the Philippines:

> Astronomers estimate that it was about 4.6 billion years ago that Earth solidified as a full-fledged planet and began to evolve toward its present-day form. The next step in this evolution was its *differentiation* into three regions known as the core, mantle, and crust. This occurred more than 4 billion years ago.²

Inevitably, this story proceeded to the forming of the first life from inorganic materials, and after its rise, how it evolved to become you.

As another example, my wife's niece is an eleventh-grade student at our local city of San Francisco (without the Golden Gate Bridge) high school, here on Mindanao. I asked her to bring home the science textbook she was currently using. As I knew it would be, it is filled with evolutionary doctrine. The three degree-holding educators who produced this 2016 edition textbook (K–12 curriculum) were trained and educated here in the Philippines.³ When dealing with origins you will read:

> How did the universe begin?...many theories have been proposed and the most accepted of them all is the big bang theory. The big bang is the event about 13.7 billion years ago when time, space, matter, and energy came into existence. This event started from a hot, dense state that has undergone

inflation–a short but rapid expansion–to become the universe that is known today.

When instructing the student on the first life you will come across this question: "How did life evolve through the course of history of Earth?" At the end of the discussion you will read this: "The Quaternary period, specifically the Holocene epoch, is the ongoing phase of Earth's history. It is in this period that humans evolved." The "necessity" of evolutionary instruction is an idea that is little more than wishful thinking, for the faithful devotees of evolutionism are unaware and oblivious to the fact they are holding to a delusion—a fantasy that is devoid of any real science. The ideas, notions, and philosophy attendant upon this concept smother anyone who happens to be born. Also, contrary to their assertion that understanding biological evolution is essential to modern biology, the pioneers of modern science and biology were by and large Christians (creationists), who believed that "*In the beginning God created the heaven and the earth*" (Genesis 1:1). Their work was neither affected nor hindered in the slightest by whether or not they believed that slime plus time turned rocks into living organisms.

In fact, it has been shown that modern science was birthed during the Reformation, for the basis of science comes from the precepts, principles, and dictums contained within the Holy Scriptures. Present-day science, with all of its benefits and advancements, arose and came about precisely when it did because the Reformation was a time when fables, superstitions, misconceptions, and irrational evolutionary beliefs and assumptions were being put to rest. These beliefs were dispelled by learned men who were first enlightened by the Spirit of God through His word.

Dr. Malcom Jeeves in his book *The Scientific Enterprise and the Christian Faith*, investigates why the Greeks never went further in their scientific studies:

> It was with the rediscovery of the Bible and of its message at the time of the Reformation...that a new impetus came

to the development of science. This new impetus, flowing together with all that was best in Greek thinking, was to produce the right mixture to detonate the chain reaction leading to the explosion of knowledge which began at the start of the scientific revolution in the sixteenth century, and which is proceeding with ever-increasing momentum today.[4]

The Reformation dealt largely with the rise of Protestant Christianity and the separation of protestant faiths from the Papacy. It was a time when the Scriptures were being re-opened, examined, studied, and proclaimed. It was then that the Christian faith of those early scientists led them to study the wonders, probe the depths of phenomena, and delve into the mysteries found throughout God's creation. These facts do not disregard, pass over, or slight the records of scientific enquiry by the ancient Greeks, Romans, and others. Nor do they discount the achievements of men like Archimedes (engineering), Euclid (mathematics), or the studies of Theophrastus in botany and mineralogy.

Various scholars have written how Christianity gave birth to modern science. Francis Schaefer in his book mentions a number of them:

> Both Alfred North Whitehead (1861–1947) and J. Robert Oppenheimer (1904–1967) have stressed that modern science was born out of the Christian world view. Whitehead was a widely respected mathematician and philosopher, and Oppenheimer, after he became director of the Instituted for Advanced Study at Princeton in 1947, wrote on a wide range of subjects related to science ... Whitehead [in his 1925 book, *Science and the Modern World*] said that Christianity is the mother of science because of "the medieval insistence on the rationality of God.[5]

Ernst Walter Mayr, an evolutionary biologist and historian of science, stated:

> Remember that in 1850 virtually all leading scientists and philosophers were Christian men. The world they inhabited had been created by God, and as the natural theologians claimed, He had instituted wise laws that brought about the perfect adaptation of all organisms to one another and to their environment.[6]

As we will discover later, those peoples who were close kinsmen of Noah and the civilizations his descendants founded shortly after the Flood, had a great deal more sophistication, scientific understanding, and expertise, courtesy of the superior intellect of their antediluvian ancestors, than evolutionists are willing to acknowledge. Down through the centuries much of the science and understanding of those ancient peoples had been lost, only to be rediscovered during the last few millennia. Wise King Solomon stated that fact this way:

> What has been will be again, what has been done will be done again; there is nothing new under the sun. Is there anything of which one can say, Look! This is something new? It was here already, long ago; it was here before our time. (Ecclesiastes 1:9–10 NIV)

Chuck Colson stated:

> ...the idea that science was invented in the seventeenth century, "when a weakened Christianity could no longer prevent it," as it is said, is false. Long before the famed physicist Isaac Newton, clergy like John of Sacrobosco, the author of *Sphere*, [sic] were doing what can be only called science. The Scholastics—Christians—not the Enlightenment, invented modern science.

Colson also quoted Rodney Stark, from Stark's book *For the Glory of God*:

> In Stark's words, "Christian theology was necessary for the rise of science." Science only happened in areas whose worldview was shaped by Christianity, that is, Europe. Many civilizations had alchemy; only Europe developed chemistry. Likewise, astrology was practiced everywhere, but only in Europe did it become astronomy.
>
> That's because Christianity depicted God as a "rational, responsive, dependable, and omnipotent being" who created a universe with a "rational, lawful, stable" structure. These beliefs uniquely led to "faith in the possibility of science."[7]

It is true that most ancient cultures used astronomy, but it was intertwined with and practiced as astrology, the ancient art of divination by consulting the stars, planets, and the signs of the Zodiac, and as such it continued up until the time of Kepler. With the rise of Christianity, the religious aspect was divorced from the mystic religious cumbrances of the Babylonian, Assyrian, and other cultures.

EVOLUTIONISM—THE STEALTH RELIGION

Unjustly maligned Creationist speaker Dr. Kent Hovind has stated that terms must be defined before one enters into a discussion on evolution; otherwise, it is futile to proceed. The following six meanings, levels, or stages of evolutionary teachings are derived from Dr. Hovind's lectures:

1. Cosmic evolution: Time/space/matter originated from nothing in the supposed "Big Bang" 13.8 billion years ago.
2. Chemical evolution: All the elements "evolved" from hydrogen.
3. Stellar evolution: Stars formed from dust clouds.
4. Organic evolution: Life formed from non-living matter.
5. Macro-evolution: Plants and animals produce offspring different than their "kind."

6. Micro-evolution: Variations develop within the kind, such as big dogs and little dogs, bacteria becoming resistant to drugs, etc.

Dr. Hovind has correctly stated that "only one of these definitions is true SCIENCE—Number 6. The first five are part of a RELIGION that adherents must 'believe' in since they have NEVER been observed or demonstrated" (emphasis his). The concept that evolutionism is a religion is an unfamiliar and outlandish thought for some, especially for those who place their unswerving faith in its doctrines. Evolutionists in reality do not worship a god. However, in one sense they are their own god: They create their own standards, their own creeds, and their own truth.

Religion for most means belief in a god, divine being, or deity. It entails a faith and devotion in service to their chosen object of affection. However, for something to be defined, delineated, and presented as a religion requires no god or worship of a deity. To be religious does not necessitate or require service to a holy being. Webster's Dictionary defines *religious* as: "relating to that which is acknowledged as ultimate reality: manifesting devotion to and reflecting the nature of the divine or that which one holds to be of ultimate importance." An evolutionist's ultimate reality and importance does not include a divine being, but *faith* in its teachings is paramount.

Without contradiction, modern science with its evolutionary foundations is a religion with its own high priests. Those are the aristocrats who hold power and influence in universities and colleges. Deans and professors hold sway over those who want to publish scientific papers, and limit access to those who are not true believers. They do not wear black robes and read from the Bible; they wear suits and ties, and their minions don white lab coats, habitually fraternize and hobnob with the remains of ancient creatures long dead, read lab reports, carry clip boards, and normally have pens and pencils in their shirt pockets. Terrence McKenna described evolutionary ideology this way:

> Modern science is based on the principle: "Give us one free miracle, and we'll explain the rest." The one free miracle is the

appearance of all the mass and energy in the universe and all the laws that govern it in a single instant from nothing.[8]

I am aware of the following complaint that is levelled at those who believe God created all things: We do not understand that the *theory of evolution* (religion of evolutionism)[9] only deals with the progression of life, which begins at that point in time when a living cell or entity produced its first wiggle or breathed its first breath. However, the religion of evolutionism is a symmetrical two-part structure consisting of the origin and formation of all material things and the evolution of life. Without both parts fully formed and in place, its scheme or system is neither complete nor fully operational. Though each part can be taught independently from the other, just as each system within the human body is independent, it is also understood that each is interdependent upon the other and only when fully combined with every other system is a human body entirely functional. Therefore evolutionism is vested with the origin of all things and comes with a comprehensive ideology and worldview.

In order to partake of points 4, 5, and 6, with its evolutionary *phylogenetic tree*[10] (tree of life), you need this tree's roots, for without its roots there would be no tree. Those roots start with numbers 1, 2, and 3, which commenced in the distant past with *nothing*. This *nothing* somehow managed to produce a *something* which *banged* in a big way, and produced all we see and observe.

On March 18, 2014, the Creation Ministries International website posted this article by Don Batten titled "Is Evolution True: 21 Bad Arguments for Biological Evolution."

> 14. The origin of life is not part of evolution. Those trained in the sciences of life, such as molecular biology, know that the origin of life is a lost cause, so some want to put it aside as "not part of evolution" because it is a gaping hole in the naturalists' argument. However, almost every major university

includes the origin of life as part of evolution in introductory biology courses. It is often called "chemical evolution." High-profile evolutionists such as Richard Dawkins agree; see the introduction to *Origin of Life*. This "dodge" is pure obfuscation because the materialist must still explain the origin of life to present a coherent view of reality, regardless of whether they can play word games with the matter.[11]

For the purposes of this book, when we discuss evolutionism we will be focusing mainly on Hovind's last three definitions.

Evolutionism insists all organisms are linked via descent from a common ancestor, and out of necessity, that original ancestor was a rock. When discussing this topic, evolutionists normally employ two terms, *macroevolution* and *microevolution*, of which only one has any basis in fact. These terms are used in numbers 5 and 6 in Hovind's list. *Macroevolution* is the transformation of one kind of living creature over a long period of time into a different organism, with different features, such as lungs, eyes, wings, or scales, not possessed by its parentage or extraction. This has never been observed and as will be demonstrated in this book, cannot happen.

The other term, *microevolution*, is a misnomer, for that is not what is commonly thought of as evolution. This process has been observed; it is simply the small changes within an animal or plant kind which have produced all the varieties we observe all around us. However, contrary to fact, evolutionists believe these minute changes are cumulative and can, in time, bring forth new kinds with features not present within the organism's ancestry. As will also be established in this book, an unlimited amount of variation or modification is never possible within any organism.

Everything has the tendency to run down, not wind up. Water never runs uphill unless it is acted upon by outside forces. What exactly were the outside forces that acted upon matter, upon the atoms and energy that existed in the beginning before time? What driving force exerted the momentum, the push, and impetus to directionless particles to organize themselves into something far greater than the sum of its own parts? How did those creatures, vegetation, and insect life cross the barriers that man has never breached, pushed over or through the limits that had never been crossed, and squeezed through the openings of doors that could never be opened? If there is no access point or open space to walk through, if something cannot be opened, it does not qualify as a doorway, does it? In reality, therefore, there were no doorways that the first primitive life could have entered or opened to get to the next level of existence.

As we will learn later, Darwin's assumption that evolution utilized natural selection to accomplish life's metamorphosis and transformation of organisms into different species, has never been found to function as assumed! It seems evolution is a concept without a mechanism.

WE HAVE A BIT OF TIME, SO LET'S GO TO THE LAB AND CREATE LIFE

Let us continue this discussion with the definition of the word *stupid*. Webster's Dictionary defines as follows: "adjective: senseless, foolish, resulting from a lack of intelligence, wanting in understanding." This book's title states the known fact that reindeer do not fly. Most people would consider someone stupid, or certainly deluded, if they truly believed they did. Conversely, this book endeavors to show why the "theory" of evolution is also a stupid dogma and system of belief. Let us now press on and see why this is so.

Those who have a somewhat tasteless, morbid, and sadistic nature, or the typical nine-year-old boy, may have asked themselves this kind of question: What does a frog in a blender really look like? Many would find it very disgusting to witness the answer. To those who entertain the idea of frog puree, the question a more cultured person might ask is, why would anyone contemplate such a thing? For those on either end of this inquiry, I have a few questions of my own.

After the blending procedure is completed on the luckless frog and the contents are uniformly mixed together, could the amphibian be reassembled and placed back together in good working order? After the reassembly, could what it lost—its shape and life-force—be reinstalled, reintroduced, programmed or breathed back into that slimy mess?

For those who have faith in evolutionism, the answer to my questions should be a resounding no-holds-barred "Yes!" If evolution created life once, without a creator, programmer, designer, or coordinator, why could not a learned man of science accomplish the same thing (maybe using different biological matter)? Surely some studied, Harvard-instructed, Ph.D. endowed scientist must be more intelligent than blind chance, lifeless water, or a rock. Of course it could be rebuilt into a new living creature—maybe not the same as the original but an animate, breathing, princess-kissing frog-type amphibian.

Evolutionary logic suggests that because life came together in the beginning from piles of uncombined, unsorted, unknown chemicals that were acted upon by undetermined forces, animating a pile of stuff that already contained all the ingredients and correct amount of required chemicals would not be impossible. Wouldn't a modern scientist who tried to create life have a distinct advantage over a mindless designer, blind chance, or random events? They would not have to start from scratch and simply guess just what kind of ingredients to throw together. Wouldn't the odds be stacked in Doctor Frankenstein's favor?

Researchers can determine the precise amount and kind of chemical compounds a living thing is composed of by analyzing some frog puree or a living creature. They also have accumulated medical and scientific knowledge from years of research and the newest, up-to-date equipment and modern laboratories at their disposal. Surely a modern lab staffed with highly trained scientists would be a better place to create life than some primordial mud puddle, watched over by blind chance and exposed to the elements and constant uncontrolled bombardment of deadly cosmic radiation.

Consider this before we continue. Any food product sitting around for a while will spoil or rot. If preserved, with the passing of time its nutritional value will decline. With the possible exception of honey,[12] all natural, unprocessed food products we buy have an expiration date. All medications and chemicals lose their potency with each passing month, thus have a limited shelf life. Gunpowder deteriorates and becomes unstable over time. Gasoline cannot be stored indefinitely—leave the fuel in your lawn-mower over the winter and see how well it performs next summer. Some things do seem to improve with age—consider certain cooked foods, which taste better when left overnight in the refrigerator and are eaten the next day—but they cannot be left indefinitely. Leave that same great tasting food in the refrigerator for a few weeks and you will want to feed it to your compost pile. And that great bottle of 50-year-old wine: How good will it taste 200 years from now?

With the passing of time, the effectiveness and usefulness of everything wanes and degenerates. Everything decays and becomes inoperative; over time, things worsen and crumble into uselessness (of course some things don't go bad, like rocks, coal, and amber, but life is made of organic compounds that do). Consider how useful, how potent, and well-preserved chemicals would be that were left sitting around for hundreds of years. Now contemplate the shelf life of those same necessary-for-life compounds just resting on the ground, lying about in the hot sun and freezing during the long cold winters of countless eons of time. Just how fresh and potent would any stockpiles of the necessary ingredients, the building blocks of life, be after sloshing around in brackish water or left lying around in the open air, exposed to the elements after a few billion years? Would not chemicals floating around in water become diluted, scattered and dispersed (less likely to be found when needed) rather than concentrated as time went on?

Before we start with the first life on this planet I must remind my readers that, before the first life cried out after its conception, the universe had to be birthed, and evolutionism's birthing agent was . . . *nothing*. That's right; they assure us *nothing* magically brought *something* into existence in a process evolutionists (a physicist coined the term and was responsible for its spread) call the *Big Bang*. But I wonder who whispered the magic word—*abracadabra*—to make that happen?

However, we will skip this part of their story, along with the formation of our solar system and creation of earth as evolutionists explain it, so that we can get right into the origin of life.

Many ask this question: What did the earliest life on Earth look like? Sam Sinai, a researcher, took a go at the answer. At the conclusion of his article he stated:

> To summarize, I think that the most primitive form of life (imaginable on earth) included elements that organized themselves to cooperate, most likely within dividing membranes. The most likely form of cooperation involved copying other molecules. Considering these, a membrane enclosing cooperative elements, that could undergo full division cycles, is the most primitive life form I could suggest.[13]

The first living thing was supposedly a simple organism—a blob-like entity, as Sinai seems to describe. Let's forget the fact that these *primitive* creatures (the theorized first life) have never been seen, nor have their remains ever been found. Simple cells, simple organisms, or creatures have never been observed, for they do not exist in the present world. Nor have any ever been discovered in the *fossil record*.[14] But this is their tale and evolutionists are sticking with it.

All living cell-based life forms have something in common: Every plant, fungi, virus, bacteria and animal contain an abundance of chemical ingredients called *nucleic acids*, which are essential for life. There are two types that reside in every organism: *DNA* (deoxyribonucleic acid) and *RNA* (ribo-nucleic acid). These coil-like strips of molecules store tons of genetic information, hold a fantastic number of blueprints, and, at the right time, issue orders to distant parts of the body to build its structures and cells. They are like microscopic computers with built-in memories, self-repairing (but defects and damage can occur that may be passed on to any offspring) and self-replicating capabilities. In the very complicated life process of each cell, the DNA contains the instructions, the RNA transcribes the DNA, and proteins are made from those instructions. Science knows nothing about any kind of cell-based life form, creature, or organism that has no RNA or DNA. Even fossils of red algae, which have been dated at 1.2 billion years—supposedly one of Earth's first lineage of multicellular organisms—has RNA and DNA.

However, organisms do exist that have neither RNA nor DNA; they simply do not exist in time nor in the present world or distant past. You will only find them lurking within the nooks and alcoves and residing securely within the dark recesses of a fantasy land, which lies within the opinion and mentality of an evolutionist.

If you are curious about this and want to know more, you need not worry; all are granted full access and may enter into their imaginary land—a land that is filled with magic and wonder, where rocks become trees and parrots over time, where slime-plus-time changes dirt into Kurt. You enter this mystic world by reading the writings of evolutionists, which usually lack the first sentence typically found in its kindred genre: "Once upon a time."

In an online resource, *Understanding Evolution: Your One-Stop Source for Information on Evolution*, there is an article titled "How Did Life Originate?" The article says the following:

> Living things (even ancient organisms like bacteria) are enormously complex. However, all this complexity did not leap fully-formed from the primordial soup. Instead life almost certainly originated in a series of small steps, each building upon the complexity that evolved previously.[15]

Wonderful short story, is it not? But is it based upon observations, studies, evidence, or science? No! Have any tests been conducted that would confirm any of those claims? Well, there are the classic chemical experiments investigating abiogenesis, which were conducted in 1952 by Stanley Miller under the supervision of Harold Urey at the University of Chicago and later at the University of California, San Diego. In his experiments, Miller, as some tell it, managed to form well over 20 different amino acids. However, his so-called primordial soup of the building blocks of life was a flask full of primordial poisons. Life is only made up of left-handed amino acids, and what was produced in Miller's apparatus was a combination of both right- and left-handed amino acids, which would be poisonous to living organisms. Unfortunately for evolutionists, life uses a set of 20 amino acids upon which to build proteins and those amino acids are always left-handed. So what Miller really proved in his experiments was this: amino acids that are formed naturally are poisonous to living things because nature does not form just left-handed amino acids but forms both in roughly the same quantities.

And yet, this failed experiment is used to prove the evolution of life. Consider this from an eleventh-grade science textbook:

> Knowing the requirements for the existence of life on a planet, scientists tried to determine how life started on Earth. In 1952, Stanley Miller and Harold Urey conducted an experiment that replicated the conditions of early Earth. *Through this experiment, they found out that life started in the oceans.* During the extreme condition of Earth, the organisms that thrived were called *extremophiles*. These organisms were able to adapt to extreme environment such as very high or very low temperatures, high acidity, or extreme pressure.
>
> As seen from the Miller-Urey experiment, the early atmosphere did not contain oxygen gas. The presence of oxygen first started when cyanobacteria emerged. Cyanobacteria were the first organisms that could produce oxygen gas through photosynthesis. But the current oxygen levels were not just due to the cyanobacteria. It was only 600 million years ago

that the oxygen levels rose to its current levels. Scientists, however, are still unsure as to what caused the oxygen levels to rise.[16] (emphasis added)

How does a failed experiment, which only produced poisonous substances, show that life started in the oceans? Then, somehow there arose at least two kinds of anaerobic organisms that began to thrive. The cyanobacteria which arose secondly, then magically acquired the ability to use photosynthesis (an amazingly complex process that man has not been able to duplicate) and produce oxygen. However, because it is also an anaerobic life form, it cannot grow in the presence of oxygen, for oxygen is toxic to it. Why would an organism develop an ability that would kill it and any other creatures alive at that time?

A BIT MORE COMPLICATED THAN WE THOUGHT

Because simple organisms have never left any evidence of their supposed existence, why would evolutionists ever use the word primitive to discuss any form of life? They assume that evolution is true, so there would necessarily be a progression of life from the simple to the more complex over time. Hence, they make their phylogenetic tree diagrams according to this baseless premise. Let us look at just one example of supposedly ancient life and see if it was truly primitive.

The Trilobite eye presents a very deep conundrum for the evolutionist. Below are a few highlights synopsized from the incredibly interesting December 2007 article by Dr. Jerry Bergman (who taught biology, genetics, chemistry, biochemistry, anthropology, geology, and microbiology at Northwest State College in Archbold, OH for over 25 years) titled "The Trilobite Eye: A Wonder of Complex Design":

- Most paleontologists believe that trilobites lived over 500 million years ago, making them one of the most ancient of known fossil groups and its eye the earliest known eye existing in the fossil record.

- Its eye lenses represent an all-time feat of function optimization, a very successful scheme of structure that is engineered with unbelievable ingenuity.

- The compound eyes of a trilobite were far more complex than ever imagined. For they were well-designed and far more complicated than the eyes of any vertebrates.

- This evidence contradicts Darwin's prediction that the earliest eyes were primitive, and that a large number of transitional forms, suggestive of eye evolution from simple to complex, would be found in the fossil record (1859).

- Trilobite scientists now conclude that trilobites possessed the most sophisticated eye lenses ever produced, and their vision may actually have been superior to current living animals.

- This "highly sophisticated" eye design is found only in the trilobite order Phacopida and is a "visual system quite different from any other eye that has ever appeared in the animal kingdom."

- Once misnamed simple primitive, a trilobite eye is now known to incorporate an incredibly complex optical chemical system into its design.[17]

When we read words in an article or story such as "a complex optical system, incredibly complex, engineered with unbelievable ingenuity, all-time feat of function optimization, and highly sophisticated," we understand the author is not describing a primitive organ or organism.

According to evolutionists, Trilobites, one of the early animals, first appeared about 540 million years ago, roamed the oceans for over 270 million years, and went extinct 240 million years ago, long before dinosaurs even came into existence. These creatures supposedly thrived during the Paleozoic Era, the first era to experience the proliferation of complex life-forms. So, if these very complex organisms existed close to the dawn of time, who are their not so complex ancestors? This information certainly contradicts the evolutionists' scenario that life has progressed from simple to complex.

Jerry Bergman's article is just about the Trilobite eye, so let's try a little more food for thought. Might the rest of this organism be even more complex? Consider this simple real-life analogy.

At one church I attended, for fundraising my family and I made peanut brittle. It was great fun and our brittle was the best I have ever tasted, bar none. In order to make the finished product, all the ingredients—the salt, the baking soda, the corn syrup, and the peanuts—had to be added at just the right time and in the proper amounts. The temperature of the mixture as it was cooking was critical. To make sure the ingredients were added at the proper time and temperature, we used a meat thermometer. This was vital, as the taste of the final product was directly related to the timing and temperature when each of the ingredients was added. Also, it was critical for the taste of the brittle when cooled, on how and when the baking soda was added to the hot liquid brittle before it was poured into the pans. Each step in the process had to be done in order, at the correct time and the proper way. Yes, there was a proper way and time to pour the liquid into the pan if you wanted good-tasting brittle. If you were off even a little during any part of the process, you would end up with some so-so or even bad-tasting brittle.

Making peanut brittle is a relatively uncomplicated process with just a few ingredients, but getting the final product right and making sure it turns out correctly is not so easy. Timing and experience are crucial if you want good-tasting peanut brittle. So, pray tell me, how would a soon-to-be-cell know the exact process, the timing, temperature, and duration of just one step? What is the name of the programmer who encoded its DNA? How would it know the precise composition of the mixture of chemicals, and in what order they needed to be combined? Remember, we are not dealing with a simple process and a simple compound like peanut brittle, but with the most complicated thing in the whole universe: life! The logical answer is that it was impossible for this supposed life-creating event to have ever just "happened."[18]

LET LIFE BEGIN

Before we begin, I have to ask a question, one I have always had. Because organic compounds and materials only come from what were once living organisms, how did the organic soup the first life emerged from come about? Well, let's just set that small detail aside and start their tale.

Imagine the process that must have been used by our first ancestors to make themselves. Pretend you are there observing this magical moment in time; you witness the whole process in time lapsed motion. How did it happen, how long did it take?

About three and a half billion years ago, on a barren, foreboding, wind-swept planet—the third rock from the sun—the first life emerged from an organic soup. All that existed on this rock before that imagined time were lifeless bodies of water, non-fertile ground littered with rocks, gravel, and an occasional volcano. The skies were dominated by cosmic rays, wind-blown dust, and poisonous and caustic volcanic fumes. If there was an atmosphere, it must have been devoid of any oxygen, for oxygen would cause any newly formed compounds to oxidize and break down long before they could be of use to a self-forming cell. However, life could not have begun without water, and there can be no water without oxygen. Besides that, without oxygen there would be no protective ozone layer, so unhindered by the lack of any kind of real atmosphere, deadly cosmic rays would have continually showered the earth and would have immediately killed any life. Yet, despite deadly ultraviolet rays, once upon a time, anaerobic life emerged within this hostile oxygen-free environment. Read what Eric McLamb teaches in his Ecology Series about the Earth's beginnings:

Earth's early atmosphere most likely resembled that of Jupiter's atmosphere, which contains hydrogen, helium, methane and ammonia, and is poisonous to humans.

As Earth began to take solid form, it had no free oxygen in its atmosphere. It was so hot that the water droplets in its atmosphere could not settle to form surface water or ice. Its first atmosphere was also so poisonous, comprised of helium and hydrogen, that nothing would have been able to survive.

Earth's second atmosphere was formed mostly from the out gassing of such volatile compounds as water vapor, carbon monoxide, methane, ammonia, nitrogen, carbon dioxide, nitrogen, hydrochloric acid and sulfur produced by the constant volcanic eruptions that besieged the Earth.[19]

Painstakingly—slowly, ever so slowly—it took millions of years for this developing life-form wannabe to gather together the necessary raw ingredients it needed—nucleic acids, proteins, enzymes, and molecules. There was no Wal-Mart around for this soon-to be cell to purchase the chemicals and other ingredients a healthy cell required for life. These compounds just happened to be there for the gathering, seemingly wished into existence, conveniently herded together—ripe for the picking. They were like a bounty of gold nuggets strategically scattered across the ground or huddled together in a glory hole, or just sloshing about in some static, briny, primordial tide pool. Although they had been lying around for numerous millennia, they remained fresh, potent, well-preserved, and had not deteriorated. Since each of the numerous kinds of ingredients, never knowing when their services would be required, continually renewed themselves so as to remain vigorous, factory fresh, and "hot out of the oven" during the lengthy interminable gathering process.

THREE CRUCIAL INGREDIENTS

Somehow, the potential cell knew just what materials to gather together to make itself. It also knew the precise amount of the building blocks of life, the correct combination and order to place them in, and the proper amount and time to add each substance to the birthing spot. By some means it knew how to set the process in motion, the formidable task that would fabricate the conglomeration of lifeless, inert chemicals to create the first living thing.

- Cartoon Studios -

However, this is not the whole story, for the three most important ingredients have not yet been mentioned. The self-forming cell also knew it needed something else, some intangible ethereal something. Because it had no mind, brain, or instinct, it is anyone's guess how it knew enough to find these elusive things and incorporate them into the tiny nest it was preparing for its birth. These certain somethings, the needed commodities, are not found in rocks, air, or the oceans, nor do they float about in the vacuum of space. They are not possessed by atoms, molecules, matter, or energy, nor the things found in nature. They cannot be seen, nor acquired from sunlight. They never occur naturally; without contradiction, they could only be formed by pre-existing intelligence. The triad of intangibles is *information*, *programming*, and *life*. Three bulwarks of scientific impossibility to cross—three hurdles of unachievable height to bound over—but our no-nothing, lifeless ancestor, whom we will call Rocky (cue the theme music to *Rocky*), rose to the challenge and conjured-up the evanescent, elusive, and shadowy components to build itself.

As far as science has determined, information never pops into existence, nor does programming arise without a programmer. The only exception might be from the wand of the tooth fairy. Otherwise, life never generates spontaneously from piles of minerals, amino acids, and chemicals left lying about. It cannot be created, cooked-up, or spawned by a mad-scientist in his castle laboratory. But Rocky, the mindless first cell, was able to accomplish these feats—through divination, no doubt.

IRREDUCIBLE COMPLEXITY

But from what source of supply did the reams of information Rocky needed arrive? Again, no plausible explanation has ever been brought forth, so the great creator named *Nothing*, must be credited with this miracle—a miracle indeed, for a cell is an irreducibly complex molecular machine with extraordinary powers and capabilities. A living cell is a brilliantly engineered, complex, intricate, and multifariously structured mechanism. It is complex in its architecture and a marvel of sophisticated, detailed design. If placed in written form, the information contained within a single cell is enough to fill a hundred million pages of the *Encyclopaedia Britannica*.

The human body contains approximately 37 trillion cells, and, as stated, each cell is an amazing machine that contains a staggering amount of information. Other than the fantastic number of these mini-cities is the amazing fact that each is unique. They are all similar, but no two are exactly alike—kind of like identical twins or triplets, only with cells you have millions and millions of identical siblings.

The amount of information needed for a one-celled organism to form fully functioning, new organs, such as kidneys, eyes, liver, or lungs, each containing millions of such cells is staggering. For someone such as a scientist or other highly educated, supposedly intelligent person to believe blind chance created those organs slowly over time and integrated them together with every other system and bodily function without a programmer is irrational; it defies reason and logic. When humans pick and choose the traits and characteristics wanted and carefully control the breeding process, there is always, with never an exception, a limit living creatures will not go beyond. No matter what is done when that point is reached, the organism will not change any further. Every attempt by man at so doing has been thwarted. There is always a natural boundary that, once reached, will admit no further change or advance. The bounds of variability have been set; they are predetermined within each organism, and cannot be crossed or ever exceeded.

Consider this analogous kind of complexity in microscopic organisms such as bacteria. Salmonella and Escherichia coli (E. coli) can propel themselves at speeds up to 15 body-lengths per second. They are endowed with a motor mechanism comprised of a rotor, stator, u-joint, hook, propeller, rods, rings, bushings, s-rings, driveshaft, outer membrane, cytoplasmic membrane, and filament. In total, there are about 40 parts to this amazing mechanism, a nano machine, comparable to an outboard or electrical motor. This motor is hardwired into sensory mechanisms or switches which get feedback from the surrounding environment. The motor runs at 100,000 cycles per minute and can stop almost instantly, within one quarter turn, then reverse and spin in the opposite direction. It is water-cooled and protein-feeding. Some describe it as the most complex and efficient engine system in the universe.

What is beyond amazing is that sundry complex machines such as these reside within each living cell. Each of these nano machines is built and assembled in proper order by other nano machines! Each machine has a different function, purpose, and task. If a person could be shrunk small enough to fit within one cell in the human body, they would see a landscape filled with astonishing phenomena—nano machines that are all programmed, fully automated, and synchronized, moving in precise fashion carrying cargo and materials to various workstations and destinations located throughout the cell.

The cell is much more complex and organized, and filled with more wonder and awe, than Dr. Seuss could have ever imagined or conceived. What is almost incomprehensible is that each machine and every component of the various machines is interdependent, a link in an extensive chain that is a necessary and vital part of the whole, and if a single part of any machine were removed, in time the whole cell mechanism would cease to run. This kind of system is called *irreducible complexity*, a term coined by biochemist Michael Behe. In his talk delivered at the American Museum of Natural History on April 23, 2002, he illustrated the term with an example of a common mousetrap:

> A common mechanical mousetrap has a number of interacting parts that all contribute to its function, and if any parts are taken away, the mousetrap does not work half as well as it used to, or a quarter as well—the mousetrap is broken. Thus it is irreducibly complex...irreducibly complex systems only acquire their function when the system is essentially completed.

THE NON-EXISTENT PROGRAMMER

Information is only useful when it is organized, structured, and arranged. For a computer or living cell, information can only be accessed and function when there is a program and the information to run the program is completely installed. No matter how much information there is, without a program, the cell or computer will not be able to utilize the data. The more complicated the machine, the more complex the program needed to run it. So who or what programmed the first cell? As the creator, named *Nothing*, gathered the information for Rocky, evolutionists again rely upon him to conjure up the software and programing needed by the still-forming, soon to be, first living entity.

Nothing guided the whole process. *Nothing* carefully arranged and gathered the data and programmed Rocky. Evolutionists also rely upon the omnipotent, omniscient, deity *Nothing* to endow Rocky with life through the possible means of some form of electrical discharge energy, such as a bolt of lightning. (Of course, lightning has never been observed in the process of creating life, and a lucky few only suffer minor injuries from its effects. However, as is the case, most of the time whenever creatures are struck by that great jolt of energy, death or harm comes swiftly to the luckless being. Now, granted, electric current is at times used to restart a heart that has stopped beating, but it is a controlled, measured jolt that is meted out. That current does not impart life to the still heart, for that mysterious force still inhabits the body. The current simply restores the proper electrical function to the switch that was turned off or interrupted, which allows the heart to once again pump the needed fluids throughout the body.)

Once this small bundle of joy came on the scene, he was not content to remain a one-celled being, for dreams of greatness and grandeur flowed through his tiny mind. Ah, the visions, hopes, and aspirations for its offspring that were concealed within this diminutive entity. He had the urge, the craving, and a pent-up desire to expand his horizons—to become all he could be. Some unexplained force, some power or compulsion drove him to evolve, to defy the laws of nature, and to acquire out of thin air new DNA necessary to change himself into a higher form of creature.

Sadly, Rocky also knew that some of his line of descent, his progeny, were fated to become food—vegetation that would feed the ever-growing number of living creatures that were destined to evolve. Nevertheless, he reveled in the vision of ponderosa pine, sequoia, and redwood trees poking their crowns into the mists and clouds to sip their moisture, pushing their roots deep into the soil to mingle and commune with their deceased ancestors, and growing to become some of the oldest, tallest, and most massive living entities on earth. For some of his offspring, he held other hopes. He wanted them to develop sophisticated brains so they would be smarter (unlike the evolutionists, whose brains currently seem to be devolving, for they are losing their intelligence, logical abilities, and reasoning capabilities, as evidenced by the belief in evolutionism). He desired that his spawn become more mobile, to become bipeds and quadrupeds and roam the vast open plains and prairies in great herds. He wanted other descendants to grow feathers and wings to conquer the vacant skies and maybe, one day, fly to the moon in machines designed and manufactured by his offspring. And at long last, Rocky desired his posterity to spread its seed out among the cosmos beyond the most distant stars.

Indulging their cravings and inner urgings, over eons of time, Rocky's ancestors crawled out of a puddle of organic broth and onto dry land. Unsatisfied with the lowly life of moss and amphibians, his descendants pushed the limits, struggled beyond impassable barriers, ever pressing upward, climbing the phylogenetic tree, branching out, and evolving into countless forms of animal and plant life.

How were these creatures—the progeny of Rocky, the beasts of the field, living and extinct, all the flora and fauna that have ever lived, able to do something no breeder, geneticist, biologist, horticulturist, or botanist, through sheer determination, diligent effort, or experimentation have ever been able to achieve—something no living creature has ever been observed in the process of accomplishing? Down through the eons of humanity's recorded history—despite their best efforts, with some of the world's most brilliant minds using the newest and most up-to-date technologies available at the time—humans have been unable to do what has been claimed to have been done by accident, by all of earth's living creatures a multitude of times.

"Do what?" you may ask. The answer: change one kind of living organism into a different kind; produce new features, different structures, and new organs with completely different functions from the parent organism or its descent family; grow or cause to form new features and new information that were not already resident within the genes or DNA of the organism, be it plant or animal, fish or fowl, amphibian or mammal; or, to put it in simple terms, over a very long period of time, change a rock into a frog and then into a prince, or change moss, algae, or mold into roses, oak trees, and herbs. The majority of intelligent and educated people would call that magic—but evolutionists call it science.

According to one evolutionist who reviewed this book, since the scientific technique of cloning was perfected a few decades ago, this became possible. Now scientists can manipulate DNA to produce animals and organs that have completely different functions from the parent organism, such as goats whose DNA was altered so they produced spider silk along with their milk in their mammary glands. However, I wonder, how healthy is it for the kids that suck from the teat, and receive milk that is infused with silk at meal time? I also ponder, out in the wild, just which kind of animal can perform those wondrous feats of manipulation to their own DNA? Even for a trained chimpanzee, I think the level of expertise needed for this technique is beyond their abilities. The technique of cloning is cutting-edge technology performed by highly trained individuals, and never happens by chance in any population of organisms. So yes, it would be magic if it happened without man's interference or manipulation of the DNA or genes.

RULE OF METHODOLOGICAL NATURALISM

Evolutionists' philosophical constraints bind them to the dictum of *The Rule of Methodological Naturalism* as evidenced by this quote: "The statements of science must invoke only *natural things and processes*. . . . The theory of evolution is one of these explanations." (emphasis added)[20] In his book *Darwin's Doubt*, Stephen Meyer states it this way:

> . . . scientists should accept as a working assumption that all features of the natural world can be explained by material causes without recourse to purposive intelligence, mind, or conscious agency.[21]

Conservapedia says that "methodological naturalism is a strategy for studying the world, by which scientists choose not to consider supernatural causes—even as a remote possibility."[22] Because evolutionists take this rule seriously, it prevents them from seeing and accepting the evidence that is there; it makes no difference how much evidence of Intelligent Design you pile in front of an evolutionist, the philosophy of Naturalism hinders him or her from even considering it.

MACROEVOLUTION AND MICROEVOLUTION

Before we conclude this chapter, as the terms macroevolution and microevolution have been used in our earlier discussion and will be used throughout this book, to dispel the notion by some evolutionists that Dr. Kent Hovind misused these words; this section would not be complete without a further examination of how scientists actually use these terminologies.

It seems these terms were coined in 1927 by the Russian entomologist Yuri Filipchenko in his German-language work, *Variabilität and Variation*. Over the years their meanings have been revised several times. Those whose religious faith holds to the philosophy of evolutionism at times blur and muddy the terms further when discussing their doctrine.

For those scientists who believe in a Divine Designer, in Chapter 2 we let Creationist speaker Dr. Kent Hovind clearly outline the meaning scientists use in respect to origins' terminology when discussing evolutionism. Of course, some evolutionists have a problem with Hovind's proper scientific definition and claim it is a misuse or a mishandling of the words, and does not accurately reflect scientific usage. Men who understand and teach science in accredited schools and universities, and who have earned degrees in the natural sciences, use the terms as explained and defined by Dr. Hovind. If that is the case, wouldn't such routine accurately reflect scientific usage? In their 1990 book *Vestigial Organs Are Fully Functional*, Jerry Bergman, Ph.D., and George Howe, Ph.D., define the terms "macroevolution" and "microevolution" in the same manner as used by Dr. Hovind.

A. E. Wilder-Smith was one of the few scientists in the world to have three earned doctorates. He was a research scientist during WWII, held professorships at numerous institutions such as the University of Geneva, was a Fellow of the Royal Society of Chemistry and a Director of Research for a Swiss pharmaceutical company. Among his many other achievements, he was also an expert on chemotherapy, organic chemistry, pharmacology, biochemistry, a gifted teacher and lecturer, and a popular public speaker. He took part in the 1986 Huxley Memorial debate at the invitation of the University of Oxford. He was author or co-author of over 70 scientific publications and more than 30 books, which were translated into 17 languages.

In Chapter 7 of his 1981 book *The Natural Sciences Know Nothing of Evolution*, Wilder-Smith stated:

> The word "evolution" is often unconsciously employed in two different senses. First, "evolution" refers to the small, often genetically hidden variations present within every species and which may be discovered in the course of breeding. Or variations may be caused by mutations and selection, which

may then be inherited (microevolution)...

Many biologists attempt to confuse the issue by using the factuality of the above type of microevolution as a basis for proving the reality of macroevolution or transformism. For this reason they confuse and intermingle these two quite distinct evolutionary concepts and attempt to blend them into one concept...Microevolution within these boundaries is, therefore, a quite different phenomenon from macroevolution, which supposedly transcends these boundaries between species or "kinds."[23]

In an article on macroevolution, the clear distinction between those two terms is casually and carelessly distorted, muddled, and blurred:

Macroevolution and microevolution describe fundamentally identical processes on different time scales...A more practical definition of the term describes it as changes occurring on geological time scales, in contrast to microevolution, which occurs on the timescale of human lifetimes. This definition reflects the spectrum between micro- and macro-evolution, whilst leaving a clear difference between the terms: because the geological record rarely has a resolution better than 10,000 years, and humans rarely live longer than 100 years, "meso-evolution" is never observed...

Within the modern synthesis school of thought, macroevolution is thought of as the compounded effects of microevolution. Thus, the distinction between micro-and macroevolution is not a fundamental one—the only difference between them is of time and scale.[24]

In other words, evolutionists conclude that both terms, macroevolution and microevolution, are names for the same processes, but operating at different scales, and that in time, both processes will produce the same outcome. This is simply an assumption, a belief, and not a proven fact.

As will be shown throughout this book, such a thing cannot happen.

Microevolution has been observed, which is simply small changes and variations in living organisms. However, these changes never produce macroevolution, which is a creature or organism transforming into a different kind of creature. Those whose religious faith persuades them to hold to the philosophy of evolutionism describe those terms in such a way as to cause confusion, which misleads the inquirer.

To sum up, speakers and lecturers on science, such as Dr. Hovind, Dr. Wilder-Smith, and others use a clear, straightforward scientific definition of the terms macro- and micro-evolution; those who espouse evolutionism use another definition that is contradictory and confusing.

NOTES ON CHAPTER 2

1. Jonathan Sarfati, Ph.D., in his 1999 book *Refuting Evolution* (Atlanta, GA: Creation), cited Dr. Stanley Jaki's book *Science and Creation*, which "documented how the scientific method was stillborn in all cultures apart from the Judeo-Christian culture of Europe. These historians point out that the basis of modern science depends on the assumption that the universe was made by a rational creator. An orderly universe makes perfect sense only if it were made by an orderly Creator" (p. 24).
2. *The New Book of Popular Science*, vol. 2 (Grolier International: Danbury, Connecticut, 1979/2000), p. 14.
3. Gloria Follosco and Adora Soriano-Pili, *Science in Today's World* (Quezon City, Philippines: Sibs Publishing House, 2006), pp. 3, 63, 65.
4. Malcom Jeeves, *The Scientific Enterprise and the Christian Faith* (Downers Grove, IL: IVP, 1971), p.13.
5. Francis Schaeffer, *How Should We Then Live?* (Old Tappan, NJ: Fleming H. Revell, 1976), p. 132.
6. Ernst Mayr, *Darwin's Influence on Modern Thought*, November 24, 2009 issue of Scientific American para. 16.
7. Chuck Colson, *Christians—Not the Enlightenment—Invented Modern Science*, CNS News web site, October 10, 2016, reprinted from the BreakPoint web site, n.d., https://www.cnsnews.com/commentary/chuck-colson/weve-been-lied-christians-not-enlightenment-invented-modern-science, para. 11.
8. Review of *The Evolutionary Mind: Conversations on Science, Imagination, and Spirit* by Terence McKenna Rupert Sheldrake, and Ralph Abraham, n.d., on Google Play (note 4).

9. A theory is based on many observations and experiments; a well-tested, verified hypothesis that fits existing data and explains how processes or events are thought to occur. As the "theory" of evolution has none of the preceding characteristics it does not qualify to be called a hypothesis, much less a theory.

    Webster's Third New International Dictionary and Seven Language Dictionary defines religion as "a cause, principle, system of tenets held with ardor, devotion, conscientiousness, and faith: a value held to be of supreme importance ... and by practicing as well as preaching its doctrines." Because in the strictest scientific sense evolution is not a theory, and is clearly a religion, throughout the course of this book, when referencing the well-known but inaccurate and flawed term *theory of evolution* I will use accurate, appropriate, and precise terminology, the much better illustrative and explanatory term *religion of evolutionism*.

10. From "Tree of life (biology)," Wikipedia.com web site, n.d., https://en.wikipedia.org/wiki/Tree_of_life_%28biology%29, para. 1.
    The *tree of life* ... is a metaphor ... used to ... describe the relationships between organisms, both living and extinct ... Its use dates back to at least the early 1800s. It was employed by Charles Darwin to express the concept of the branching divergence of varieties and then species in a process of common descent from ancestors ... The modern development of this idea is called the phylogenetic tree.

11. Don Batten, "Is Evolution True: 21 Bad Arguments for Biological Evolution," Creation Ministries International web site, March 18, 2014, https://creation.com/is-evolution-true, para. 18.

12. Honey was engineered by God to be stored and eaten by bees during times of food scarcity. It really is true: God designed this food so well, that if handled and stored properly, honey does not spoil and seems to have an indefinite shelf life. It is said that sealed honeycombs from King Tut's 3,000-year-old tomb were found to have edible honey inside them, but I wonder who was brave enough to taste honey that was covered in mummy dust in order to verify that fact.

13. Sam Sinai, [Response to the question] "What was the first form of life to form on Earth? How was it formed?" Quora web site, October 10, 2016, https://www.quora.com/What-was-the-first-form-of-life-to-form-on-Earth-How-was-it-formed.

14. The so-called "fossil record" should be properly called the *fossil graveyard*. The fossils are a record of death. They do not tell a story of relationship, ancestry, or the succession of life between once living creatures, but their demise, which happened within a few days, weeks, and months of each other.

15. University of California Berkeley, "How Did Life Originate?" *Understanding Evolution* web site, n.d., http://evolution.berkeley.edu/evolibrary/article/origsoflife_04, para. 1.

16. Follosco and Soriano-Pili, 2006, p. 10.
17. Jerry Bergman, "The Trilobite Eye: A Wonder of Complex Design," Creation Science Association of Alberta web site, December 1, 2007, http://www.create.ab.ca/the-trilobite-eye-a-wonder-of-complex-design/, paras. 1,
18. Adapted from Michael Earl Riemer's, *Musings on Creation and Evolution* (Indianapolis, IN: Dog Ear Publishing, 2014), p. 27.
19. Eric McLamb, "Earth's Beginnings: The Origins of Life," Ecology Global Network web site, September 10, 2011, http:// www. ecology. com/2011/09/10/earths-beginnings-origins-life/, paras. 9–11.
20. National Academy Press, Teaching About Evolution and the Nature of Science, 1998, pg. 42.
21. Stephen Meyer, quoted in Paul Nelson, "Methodological Naturalism: A Rule That No One Needs or Obeys," Evolution News and Science Today web site, September 22, 2014, https://evolutionnews. org/2014/09/ methodological_1/, para. 3.
22. "Methodological naturalism," Conservapedia web site, May 19, 2017, http://www.conservapedia.com/Methodological_naturalism, para. 1.
23. A. E. Wilder-Smith, *The Natural Sciences Know Nothing of Evolution* (Green Forest, AR: Master Books, 1981), pp. 123–4.
24. "Macroevolution," Project Gutenberg Self-Publishing Press web site, n.d., self.gutenberg.org/articles/eng/Macroevolution.

> The evolution of the genetic machinery is the step for which there are no laboratory models; hence one can speculate endlessly, unfettered by inconvenient facts, ...
>
> We can only imagine what probably existed, and our imagination so far has not been very helpful.[*]

CHAPTER 3
THAT MYSTERIOUS SOMETHING[1]

IT IS THE ESSENCE of being, an elusive mysterious something, sometimes called energy, a vital force, a property, animate existence, a shaping force, or a principle. There is no universal agreement as to its definition,[2] for it is a force that defies precise description and explanation. It is something that can be sustained for a few moments, may last for decades, and in some cases for thousands of years. It cannot be created, duplicated, produced, fabricated, or manufactured by man. It is that thing which distinguishes inorganic matter from organic matter. It never arises spontaneously from non-living compounds, materials, or inorganic matter. It is something that every kind of species that has ever existed possesses and passes it along, one generation to the next. It has long been studied, mused over, pondered, and reflected upon; it is called *life*, "a phenomenon almost impossible to define or explain in all of its varying aspects."[3]

---

*Richard E. Dickerson, Ph.D. (physical chemistry) (Professor, California Institute of Technology), 'Chemical evolution and the origin of life'. *Scientific American*, vol. 239(3), September 1978, pp. 77 and 78

Life

That mysterious force, the essence of being,
Existence, subsistence, God's power you're seeing.
Your vigor and spirit, the zest and the verve,
A living being means God's somewhere near.

He's upholding, sustaining all things within sight
And things that we don't know that pass in the night.
With power and skill, puissance, and brawn,
God's mercy and love in His beings is shown.

Dirt by itself inert it will be;
Inactive and static, no movement you'll see.
And stones do not grow, a fool you would be
To think that they'll eat or grow into a tree.

A creature deceased, devoid of all life,
Cannot be brought back, will never live twice.
It can't be revived, grows cold just like ice.
Only the power of God can give life.

All wisdom and power and strength to the Lord.
All honor and glory, He gives life by His Word.
All beings does He cause to exist.
Only fools refuse to believe all this.

In this chapter we will be exploring and delving into two views of reality. Scripture proclaims that God is the author and Creator of all things. "*In the beginning God created the heaven and the earth*" (Genesis 1:1). Evolutionism proclaims: In the beginning nothing created all things out of nothing.

In 1818, twenty-year-old Mary Wollstonecraft Shelley published a novel entitled *Frankenstein; or, The Modern Prometheus*. The protagonist was a character by the name of Victor Frankenstein. He was a scientist, who as a boy longed to discover the fabled *elixir of life*. He became obsessed with the idea of creating life in inanimate matter through artificial means. After a time of studying chemical processes and the decay of living creatures, he reasoned that he had gained enough insight, so much so that he attempted to create life. He assembled a grotesque, but sapient humanoid creature from parts of dead bodies and successfully brought it to life through artificial means.

Although just a fable, the belief that life can be generated from nonliving matter is very much alive within the hearts and minds of those who hold to the philosophy of evolutionism. To them there is no *vital force*, or as the *Encyclopaedia Britannica*[4] puts it, "a kind of ghostly mainspring"; they hold that life results from just the right combination of chemicals coming together in the correct proportions, and that given the right set of circumstances, maybe a lightning bolt out of the blue striking a preassembled pile of chemical compounds, shazaam, abracadabra, presto-chango, life can be kick-started—just what Doctor Frankenstein accomplished.

Mary Shelley's Doctor Frankenstein and his monster did not exist. However, even today, with all our modern technology and understanding of medicine, we are no closer to really understanding what life is than we were in 1818. Regarding the mechanics, the functions, and such of life, of course we now know a great deal more; however, just what life *is* is still a great mystery. To evolutionists, however, there really is no great mystery. They believe that one time in the distant past, somehow a bunch of chemicals combined themselves together in the proper amounts, and when touched by an electric current, life sprang forth.

Shelley's Doc Frankenstein, unlike the modern evolutionists, did not start from scratch with his experiments. He used pre-existing organic materials. Today, however, if some mad doc with the urge to create life had fresh flesh to work with, the advantage of knowing the right combination of chemicals in the correct proportions that was all prewired; he would still not be able to bring back to life that which was once alive. Whether life was slowly drained out of a body through a debilitating disease, or is suddenly ended by an accident or heart attack, once life is gone nothing will bring that person back from their ultimate fate. At the moment of death the soul/spirit leaves the body and goes to the place prepared; heaven for the righteous, and eternal punishment for the wicked. No living creature whose soul/life has left its body has ever been brought back to life without the intervention of the Creator of Life: God. Why even carve up dead bodies and sew them together when there are plenty of whole bodies that would be available for reanimation? I am sure there are a few billionaires who would pay handsomely if they believed it possible; to reanimate a recently departed loved one, or even themselves.

At least one evolutionist who reviewed this book disagreed with me on the point that once life has left the body, nothing will bring that person back from the dead. She referenced two cases she believes invalidate my proclamation, those of Paulie Hynek and Justin Smith. She stated: "In those instances, life was in fact sustained and resumed, respectively, through artificial means." In both cases the individuals had accidentally passed out and wound up spending an entire night in a snow bank. In 2001, by the time 2-year-old Paulie Hynek of Eau Claire, Wisconsin was found, his body temperature had dropped to 60 degrees, his heart had stopped beating, and he was no longer breathing. There were no signs of life and his body was frozen to the point that he was literally getting stiff. In the other case, Justin Smith's body temperature when found was just a bit warmer, but still a bone-chilling 68 degrees. Both got as close to death as is seemingly possible and still have a spark of life within. They were both revived through the skill of the attending physicians and hospital staff. In these cases life was not created, a corpse was not resuscitated, for life was still there, very faint, weak, like a dying ember, but as Mark Hynek, Paulie's father related to journalist Mike Nichols: "...each of us has electricity in the body. He thinks of it as light, sort of like the lights that shine in the sky at night in 'It's A Wonderful Life,' and you're not totally gone, he says, until that light flickers and goes out."[5]

Even though life cannot be created, there are people who still believe reanimation is possible and have taken steps to be brought back from the dead at some future date; this process—or would-be industry—is called cryonics. Those who believe it will work hope that in the future a real Doctor Frankenstein will emerge to rescue them from their fate (the belief in spontaneous generation really does die hard, doesn't it?).

CRYONICS

The word cryonics comes from the Greek κρύος (*kryos*) and means "icy cold." Wikipedia, the free encyclopedia, states that cryonics "is the low-temperature preservation of humans and animals that cannot be sustained by contemporary medicine, with the hope that healing and resuscitation may be possible in the future." Although most scientists regard cryonics with skepticism, a few (at least 68 from the Evidence-Based Cryonics organization), supported the idea in a letter (to anyone who was interested) titled "Scientists' Open Letter on Cryonics."[6] The procedure was first proposed in 1962; by 2012, approximately 250 people had chosen to undergo the procedure. Although the process and maintenance of this procedure is somewhat expensive, the cost has not deterred the faithful.

According to reporter Zack Guzman's April 26, 2016 story *Who Wants to Live Forever?*,[7] in Scottsdale, Arizona, for the small price of just $200,000, you can have your whole body popsicleized by liquid nitrogen after death, in the hopes that future medicine might be able to raise the dead. Right now, frozen in liquid nitrogen, there are at least 147 heads and bodies, waiting and hoping one day to be revived. You read that right, *heads* and bodies. For those like Elaine Walker, a 47-year-old part-time college instructor who could not cough up the $200,000 necessary to revive her dead carcass, but for only $80,000, she is going to have (presumably, after her death), a surgeon remove her head, in the hopes that at reawakening-time a new body can be grown from her tissues' DNA.

I think I will forgo becoming a popsicle, and just let Jesus call my name after my demise, for the new body He will grow for me will be much preferable to any home-made, man-grown body. Have you ever had a brain freeze after eating ice-cream? Imagine the kind of pain a brain freeze resulting from your head thawing out after being frozen solid by liquid nitrogen; I shudder and tremble when I contemplate such a horrible aftereffect.

An even greater concern might be what happens if a new body cannot be grown after your head is thawed out? Would you end up like Professor Donald Kessler (Pierce Brosnan) in *Mars Attacks!*, a 1996 black comedy science fiction film directed by Tim Burton? In the film, the professor is captured by Martians, who dismember his body but are able to keep his disembodied head animated, fully aware, and able to talk. Well, be positive, there is good in everything. If you ended up like that, you would certainly save money on food and clothing. I just love the quote in Guzman's story from Columbia neuroscientist Dr. Ken Miller, who likened cryonics to "selling tickets to a ride you can't go on."[8]

Just think about it. If evolution is possible, why would reanimation not work (of course many believe that cryonics at this time is not possible)? For as evolutionists believe, in the beginning piles of non-combined chemicals, all the necessary amino acids and the whole shebang, just waited for that proper lightning strike to animate that lifeless stuff into a wonderfully complex, fully functioning cell. Now along comes modern Doc Frank, and everything he needs is lying on a table in front of him: a corpse (or, in Elaine Walker's case, just a head), which is something that has everything already preassembled, just waiting for him to imbue it with life (and grow a body if needed).

Supposing evolution to be true, to create life all we would need are two things (three, if you count the lightning strike or electric current). First, a modern Doc Frankenstein who knows a great deal of modern medical science and is willing to try to do the impossible, and second, a young man or woman that has recently died a non-violent type of death, such as hypothermia, which would not horrendously damage a body. So, for a moment let us step into the life of a modern re-animator and watch as he attempts the impossible.

DOCTOR HULBERT

Doctor Hulbert, a modern Frankenstein wannabe, has located his laboratory in the midst of a large metropolitan area, with the reasoning that it would be much easier to obtain the needed organic materials when surrounded by a vast supply of potential future subjects. He also knows that time is of the essence, and after attaining a fresh corpse, precious time would be lost if it had to be transported some distance out to the middle of nowhere. His modern laboratory has been outfitted with all the old standard equipment, as well as the newest experimental medical apparatus. His research and equipment was partly funded from the generous contributions of a few billionaire investors who are very interested in having access and a stake in his reanimation techniques. The remaining funding came from the coffers of a billion-dollar, multinational, pharmaceutical conglomerate whose CEO and other executive officers were intrigued by his lectures and seemingly logical explanations as to how and why he could pull it off. Thinking of future profits to be made, they dipped deeply into their vast off-the-books expense accounts and supplied the doctor with a large, eight-figure, unpublicized grant.

The doctor's assistant is not the iconic Igor; the dim-witted hunchback oaf often portrayed in horror movies as an assistant to Doctor Frankenstein (but interestingly, was not a character in Shelley's original book). Instead, the doctor's assistant is the young and confident Mark Doubleday. After carefully evaluating a host of potential candidates, he was by far the most gifted and qualified, far superior to the rest. He was a recent graduate of the top-ranked Johns Hopkins University, where he had earned two PhDs in the field of biomedical sciences. He was also the youngest professor in recent memory at the school in which he was formally a student. Professor Doubleday was much taken with the elder, brash, and confident scientist. Most importantly, he knew that with his help the doc would succeed.

With his funding all secured, all necessary permits obtained (from public officials who looked the other way and were given generous contributions for their political aspirations), his laboratory all equipped and prepped, an able assistant standing by, all he had to wait for was a fresh test subject who voluntarily or involuntarily donated his or her body to the doctor's noble cause.

He does not have to wait long. Directly after completing the final installation of the remaining equipment, the sound of an emergency vehicle can be heard coming from a distance. Its siren grows louder as it approaches the loading dock. As the brakes lock, the tires screech and the vehicle comes to a jolting stop. The doctors hurriedly unload their first test subject.

As Doctor Hulbert makes his final calculations, the professor prepares the body. Wires are placed strategically in precise locations and intravenous drip tubes are inserted into the corpse. As the old blood is being drained, they must match its type against the in-house supplies (every major type is available in their vast storehouse) before the fresh blood serum can be pumped into the corpse.

When all is prepared, Doctor Hulbert's blood serum, which contains his elixir of life, a special mixture of nutrients and chemicals he has designed during years of painstaking research and experimentation, is warmed and pumped into the waiting subject. Everything is going according to the well-rehearsed and planned procedure.

The duo work like men possessed, laboring over the corpse lying on the table. At last the time they had been waiting for, the event that would give either great jubilation, or crushing defeat had come ... the electric current, the giver of life, must be applied. Jolting the corpse with flowing electric current, using different amperage and voltage for longer and shorter durations of time, they start and stop the current over and over again. The corpse twitches and spasms each time the current is applied, but that is it, nothing else happens. After a while they see what looks like smoke arising from the corpse, and a distinct smell of burning flesh emanates from the body lying on the table. Undeterred, Doc Hulbert finally says, "That's it! Away with this stiff and get me another one." This goes on day after day, month after month, until they have plied their wiles upon enough lifeless bodies to populate an entire small city. They say, "Never give up!" However, it is also said, "There is no point in flogging a dead horse."

Let us now leave the dream-world of Doctor Hulbert, Professor Doubleday, and all the modern-day evolutionists. We must face the reality that, even with a pre-assembled, properly wired corpse with all its parts in their apt order and place, nothing we can do to it will re-endow it with that elusive, mysterious something called *life*. That light switch that allows the current (life) to flow through an organic being can never be created in a laboratory or anywhere else outside nature, for when something dies that life-force is now once again back within the hands of the one who created it in the first place—*God*—and He is not about to hand it over to a deluded evolutionist.

The true answer to where the mysterious something comes from does not lie with the god of the evolutionists, *Nothing*, but is contained within the pages of the Holy Scriptures' first few chapters:

> In the beginning, God created the heaven and the earth. (Genesis 1:1)
>
> And God said, 'Let the waters bring forth abundantly the moving creature that hath life' (Genesis 1:20)
>
> And the LORD God formed man of the dust of the ground, and breathed into his nostrils the breath of life; and man became a living soul. (Genesis 2:7)

NOTES ON CHAPTER 3

1. This chapter originally appeared in Chapter 13 of Riemer's (2014) book *Musings on Creation and Evolution*, and has been revised for this book.
2. "There is no generally accepted definition of life," *Encyclopaedia Britannica Micropaedia* (1943–1973, 1977) vol. 10, p. 893d.
3. William Benton, "*Life*," *Encyclopaedia Britannica Micropaedia* (1943–1973, 1977) vol. VI, p. 212.
4. Benton, "Life," p. 894.

5. Mike Nichols, "How Paulie Hynek Died One Night, and Came to Live Again," Badger Diggings [blog], JS Archive, December 2006.
6. *Scientists' Open Letter on Cryonics*, n.d., Evidence-Based Cryonics web site, www.evidencebasedcryonics.org/scientists-open-letter-on-cryonics/.
7. Zack Guzman, "Who Wants to Live Forever? $200,000 for a Shot at Immortality," CNBC web site, April 26, 2016, https://www.cnbc.com/2016/04/26/meet-the-company-offering-a-chance-at-immortality-for-200000.html.
8. Ibid., para. 19.

The family tree of the horse is beautiful and continuous only in the textbooks. In the reality provided by the results of research it is put together from three parts, of which only the last can be described as including horses. The forms of the first part are just as much little horses as the present-day damans are horses. The construction of the whole Cenozoic family tree of the horse is therefore a very artificial one, since it is put together from non-equivalent parts, and cannot therefore be a continuous transformation series.[*]

CHAPTER 4
WALKING AMONG HORSE-AND PUMPKIN-KIND

Jamal was an award-winning architect and an outstanding engineer. He was supervising the construction and design of a sizable building project with an ocean-based color theme. He hoped that, when completed, the tones of aqua blue would invoke feelings within a person's mind and moods, and that the ambiance of the building's rooms and qualities would create a certain atmosphere—an impression of peace and tranquility.

Because Jamal had a thorough knowledge of and background in interior decoration, architectural design, and ceramics, he planned to mix all the colors and make all the tiles and glaze onsite. He also thought it

---
[*]Prof. Heribert Nilsson, *Synthetische Artbildung*, Verlag CWE Gleerup, Lund, Sweden, 1954, pp. 551-552.

best to add the pigments to the paint, plaster, and cement during the preparation instead of painting or applying a layer of pigment over the finished pieces. He did not need to keep a cornucopia of all the various shades of color on hand; he needed only the three primary colors—red, yellow, and blue—plus black and white to have all of the colors of the rainbow available to him.

It is an interesting fact about the three primary colors that they are the only colors that cannot be made by mixing two other primary colors together. By mixing various amounts of black and white with any two primary colors, you can create a multitude of tones, tints, hues, and shades of those primary colors. But if you lack one of the primary colors, it matters not the length of time you spend mixing, sorting, combining, or blending the other two primary colors; certain hues, shades, and colors will be impossible to make.

Now consider the original genus, kind, or category from which the first dog, cat, bear, rabbit, or any of the oodles of other creatures sprung. Residing within the unique DNA warehouse of each class of animal was all the information—all the primary colors—available to produce every trait, feature, and type of each animal we observe all around us. The supply of information was packed within the DNA of that first kind, the parent of its line of descent. However, that available supply can never produce any traits or varieties other than those which have been pre-determined by the primary colors. Now, whether you are considering the question of the origins of life from the perspective of evolution or from belief in a creator, the information needed to produce each kind or species' line of descent resides within the first of its kind.

One reader questioned my line of reasoning and pointed out examples of creatures such as the Vacanti mouse that supposedly disprove my assertion. The mouse was named after researchers Joseph and Charles Vacanti who, searching for a way to grow cartilage structure for surgeons to use to reconstruct human ears, managed to grow ear-shaped cartilage on the back of a laboratory mouse. The "nude mouse" used was not genetically engineered, but was the result of a spontaneous natural genetic mutation that left it without hair and virtually no immune system. The ear-shaped cartilage structure was grown by seeding cow cartilage cells into a biodegradable ear-shaped mold. After implanting this under the skin of the mouse, the cartilage grew by itself. Genetic defects such as occurred in the mouse happen once in a while: We have all read stories and seen pictures of blue crayfish, two-headed snakes, and other oddities nature serves up. But when this happens it is a corruption of the genic code or DNA, and in almost all cases cause harm to the individual organisms that happens to be afflicted. As far as the "ear" that grew on the back of the mouse, this would not happen in the wild, for there are no animals that I am aware of that can perform these kinds of experiments and procedures.

However, if, as evolutionists believe, there really is no such thing as an original kind of anything—no original plant or animal—other than the archetype rock material from which all living things were sprung, thus, there would be no static genus nor any distinct kind; there would be no limits, no boundaries, and no delineation or demarcation point of reference to determine one species or kind from any other. Every living thing would be relentless in its perpetual endeavor—its constant metamorphosis—to become something else.

Nevertheless, as we observe,[1] each kind of plant and animal has its own special genes, unique DNA, and certain primary colors—colors that are neither genetically comparable nor compatible with those of any other kind. They are unique to that species and will neither work nor function in any other kind. This is why, for example, if you are attempting to hybridize bananas in order to produce another variety of that fruit, your efforts will never magically, nor accidently, yield bananas with feathers, scales, feet, or a pair of eyes (unless man messes with the DNA or genic structure). The primary colors—the genes or DNA needed to fabricate those traits—do not exist within the banana.

LARGE AND SMALL CANINES

Speciation, such as that which is done in the breeding of new dog varieties, does take place; however, contrary to the logic of evolutionists, its occurrence—whether by man's interference or nature's design—is not evolution in action. Speciation can never produce an entirely different species or kind of animal; it can never create a rat from a dog. With each new variety or strain, there is a loss of primary color information that reduces the supply of available colors. It does not increase the amount but simply alters the kind of genetic material available.

Witness the end of such selected breeding in the miniature and toy poodles, or consider Chihuahuas or other miniature dogs, such as the Chinese Crested. These small breeds are prone to genetic anomalies—often presenting neurologically, as conditions such as epilepsy and seizure disorders, and in physical deformities, especially in old age. At the other end of the size spectrum, large breeds of dogs such as the Great Dane, Mastiff, Dogue de Bordeaux, and Newfoundland, often possess tender and loving temperaments, but they are also prone to physical deformities and other genetic anomalies. These four-legged giants can certainly perform such tasks as pulling carts with heavy loads, guarding the family, property, flocks, and hunting big game, but there are other factors to consider here, such as veterinary bills, which can be enormous due in large part to the selective breeding process. With both ends of the dog gamut, you could put it this way: These dogs are a few genes shy of a full load.

While many types of dogs serve a useful or necessary propose, the many genetic defects and other anomalies would render over-bred animals incapable of surviving if left to their own devices out in the wild. Natural selection would do it its job and select them for termination.

MINIATURE AND DRAFT BREED HORSES

Let us now further contemplate a few more examples of the limits of variation in the animal kingdom. There are two extreme sizes in the horse family that breeders have been able to produce: Extra-large and super-small.

The smallest female horse in the world, officially recognized by Guinness World Records, is a dwarf miniature named Thumbelina. Like many dwarf animals, she is stout and possesses short limbs. She was born in St. Louis, Missouri on May 1, 2001. She stands just 17" tall and weighs in at a petite 57 lbs. Another record holder named Einstein, at birth was just 6 lbs. and 14" at the withers and was the smallest horse on record to ever survive. The remarkable thing about this miniature horse is that he shows no dwarf characteristics, he is simply an extremely small miniature horse. In April 2010 this stallion entered Guinness Book of Records. However, Einstein's title did not stand for long. Bartolomeo Messian of Italy is now the owner of the smallest living stallion. Charly, an Arragon Arabian, measures just 25" to the withers. He was born in Holland in 2007 and imported to Italy. He took the title in April 2012.

The Tinker "Mario" and the world's smallest show horse and mini-horse "Charlie" Photo: Horst Galuschka/dpa | usage worldwide Contributor: dpa picture alliance / Alamy Stock Photo

You would not see a herd of these miniature or dwarf horses mingling with herds of normal-size horses, for they would get trampled underfoot. Nor will you see them running lose in the wild like mustangs; they are so small they could probably be taken down by a large house cat or hefty shrew. Have the breeders reached the diminutive size limit with Charly the stallion? Or is it possible to produce an even smaller horse? If there are no limits, a full-grown horse the size of a mouse could someday be bred. And as they kept at their craft and continued to push the thresholds of biology, an ant-sized equestrian could be brought into existence. Of course, finding a saddle and a jockey to ride a horse that size will still be a big problem. Charly is probably at or close to the absolute size limitations for miniature horses. On the opposite extreme of the miniature horse stand the giant, heavy draft breeds, such as the Shire, Clydesdale, Suffolk Punch, and the Belgian. Many weigh in at nearly a ton and can stand over 6 ft tall.

Traditionally, horses and ponies have been measured in units called hands. One hand is equal to 4". To determine the height of a horse, you do not measure to the top of its head, but you start from the ground just beside and behind a front foreleg, and measure to the top of the fifth vertebrae or withers. This is the only place on a horse that does not change in height, whether he drops or arches his back or lowers or raises his head. The neck and head of a horse generally extend in height another two or three feet above the withers. When standing on the ground in front of the average tall horse, most people would neither be able to touch the top of its head with their hand nor look at them in the eye.

In 1850 a giant Shire Gelding named Samson (renamed Mammoth when he was four) stood at 21.2½ hands (86.5") and weighed an astounding 3,360 lbs, making him the largest horse to ever live in the history of large horses. More recently, a Shire gelding named Goliath, at 76" tall, was the Guinness Book of World Records record holder until his death in 2001. Another tall horse, a resident of Australia, is Noddy, also a Shire. When measured by his owners in 2008, he was found to be over 80" tall. His owner Jane Greenman said this about him: "It sounds like a mountain moving when he gallops across the paddock to come and get his breakfast... He eats an incredible amount. I would hate to run a team of eight horses his size—it would send you broke."[2]

As of April 10, 2014, Jerry Gilbert of Smokey Hollow Farm in Poynette, Wisconsin, USA owns the tallest horse in the world. He is a thirteen-year-old Belgian gelding named Big Jake. He is certified by the Guinness Book of World Records as the world's tallest horse. He stands 20.3 hands or nearly 6-foot-11 and weighs in around 2,600 lbs. He also has an enormous appetite; he can munch through 1½ bales of hay and 40 quarts of oats in a single day.

Using just the draft breeds such as the Shire, Belgian, or Clydesdale as a starting point, would it be possible to produce a miniature horse line? Without mixing in any other type of horse, wild or domesticated, could a miniature stallion like Charly be bred? Or if a breeder started with the miniature horse blood line and selected only similar kinds, without adding or using any other type, could he eventually produce or attain a draft breed Shire or Clydesdale type of horse?

It is highly improbable and most likely impossible that using only draft horse lineage or stock, could a breeder produce a miniature strain of horse. Likewise, using only miniature or dwarf parentage and not mixing in any other type, it is highly unlikely that a large draft horse could be attained. The reason for this is that the information needed to produce another breed no longer exists within their respective gene pool, and there is no known artificial or natural way to create the needed information. Notice I wrote *create the information*, not transplant, gather, or harvest it from some already existing organism. All the information needed to produce all the varieties of horses that now exist, at one time existed within the original horse kind. But down through the ages, nature at times, but for the most part mankind, selected the traits and characteristics they found desirable and bred out the unwanted ones (information).

Through the selective breeding process, would it ever be possible for man to produce a pure white stallion with fully functioning wings that allowed it to fly like Pegasus, the horse of Greek mythology fame?

But why stop with wings? Why not develop a winged horse with aquatic characteristics such as gills, scales, and fins? Then it would be able to swim the depths in search of succulent sea grasses to graze upon, and when finished dining, ascend to the clouds and soar among the eagles. Are those ideas more preposterous than the belief that a rock evolved into a horse? Are those thoughts really any different than believing that intelligent biologists, geneticists, or scientists guided by reason and purpose could one day develop those kinds of horses, or than believing that blind chance—with no direction, no guidance, no intelligence, or purpose—after a very long, drawn-out process turned a rock into a magnificent mustang, bald eagle, blue whale, hummingbird, or impressive redwood tree?

In every kind of animal or plant ever produced, groomed, bred, or grown, it has never been possible to increase the amount of information inherent within the organism. All that has ever taken place is a reshuffling, a rearrangement of existing programing. A loss of information is always the result; never is there an increase in information within the organism that would be needed to form a new feature or organ in a living being.

What evolutionists never seem to consider is that you can crossbreed members of the horse kind, dog kind, cat kind, or any kind, for as long as you want, but you will never get anything other than the kind you originally started with. Each new variety will have less genetic information than when the selective breeding process started, which is just the opposite of what evolutionism requires. Because nature cannot create information out of thin air, nor has mankind as of yet figured out a way to fabricate information and insert it within the genes of a living creature, how has evolution managed to perform this impossible feat innumerable times?

Of course, an evolutionist might say that given the recent advances in genetic editing, we may indeed have a way to fabricate new animals. That may well be, but I don't think we will see wild animals getting together anytime soon to practice their gene editing techniques.

THE PUMPKIN KIND

Let us now consider the limits of growth and variation for a part of the plant kingdom: the pumpkin, which botanists and *Webster's New World Dictionary* define as a fruit, next-of-kin to tomatoes, beans, and green peppers. For the rest of us, classifying beans and green peppers as fruit seems a little odd; we call those plants vegetables. Here is a thought for botanists: if you mixed pumpkins, beans, green peppers, and tomatoes together and bake them in a pie crust, would you put whipped-cream and a cherry on top, call it a fruit pie and eat it for desert? If someone were to ask me whether a pumpkin is a vegetable or a fruit, my answer would be, "Yes."

At the 1893 World's Fair in Chicago, William Warnock, a farmer and machinist from Ontario, was the first to win a world record in North America for a giant pumpkin, a 400 lb oversized fruit. In 1903 Warnock again made history at the St. Louis World's Fair with a 403 lb pumpkin. That pumpkin held the world record for more than 70 years. In 1981, Howard Dill made it into the Guinness Book of World Records with a mere 493.5 lb pumpkin. For many years, growers tried to break the 1,000 lb barrier. The record finally fell twice in 1996, with weights of 1006 and 1061 lbs. Since then, nearly every year a new world record weight is announced for the orange gourd family.

Back in 2006 Ron Wallace from Greene, Rhode Island, broke the world record with a substantial 1,502-pounder. A few years later, on September 28, 2012, Ron brought his record-breaking 2,009 lb pumpkin to the Topsfield Fair in Topsfield, MA. At the Half Moon Bay Art and Pumpkin Festival, held in Californian in 2014, John Hawkley won the prize for the largest pumpkin grown in the U.S. It was a monster, weighing in at 2,058 lbs. However, the heaviest pumpkin grown in the world that year was by Swiss gardener Beni Meier (age 30). To transport his trophy pumpkin to the Great Pumpkin Commonwealth Association for weighing, he had to use a special vehicle. When weighed, it tipped the scales at an impressive 2,096 lbs—a world record. However, within a few weeks he had two other pumpkins tip the scales at 2,102 and 2,323 lbs respectively!

Mathias Willemijns poses with his world record (1190.5 kg) pumpkin at the European Championship Pumpkin Weigh-Off in Ludwigsburg, Germany, October 9, 2016. Photo: Christoph Schmidt/dpa/Alamy Live News Contributor: dpa picture alliance / Alamy Stock Photo

In 2015 Ron Wallace from Rhode Island again clinched the U.S. title with a 2,230 lb gourd. In 2016 Richard Wallace, Ron's father, grabbed the title from his son when he took home the top spot at the Frerich's Farm Pumpkin Weigh-Off, the most prestigious pumpkin growing competition in America, with a gourd weighing in at 2,261.5 lbs. But even that monster of a pumpkin was no match for the King Kong of fruits, to date the heaviest pumpkin ever grown. Weighed and certified at the Giant Pumpkin European Championship held in Ludwigsburg, Germany, the champion, grown by Mathias Willemijns of Belgium, was an astounding 2,623.5 lbs!

Every year the race is on to see just how big the growers are able to coax the delicious pie fruit. It is obvious that the limit of its size has not been reached, for pumpkins continue to pack on more pounds each year. What is the size and weight limit? Is a 5,000 lb behemoth, with a 30-foot girth, within the realm of reason? Can it grow to the size of a small house or as large as a football field? Somewhere between the bulk of the current record holder and Mount Everest, the limit will be reached.

There is a natural limit to growth and variation programed into the DNA of all living things. Sooner or later, pumpkin growers will reach the size and weight limit, and once that point is attained, no matter what they do, no one will not be able to grow a pumpkin one ounce over that limit. Unlimited change, variation, and transformation are never possible, despite what the evolutionists' phylogenetic tree chart teaches. I find the following humorous tale comparable to the reasonings, logic, and conclusions reached by those "scientists" who hold to their belief in evolutionism.

FELIX THE FLYING FROG: A PARABLE ABOUT SCHEDULES, CYCLE TIMES, AND SHAPING NEW BEHAVIORS

Once upon a time, there lived a man named Clarence who had a pet frog named Felix. Clarence lived a modestly comfortable existence on what he earned working at Wal-Mart, but he always dreamed of being rich.

"Felix!" he exclaimed one day, "We're going to be rich! I'm going to teach you how to fly!"

Felix, of course, was terrified at the prospect: "I can't fly, you idiot... I'm a frog, not a canary!"

Clarence, disappointed at the initial reaction, told Felix: "That negative attitude of yours could be a real problem. I'm sending you to class."

So Felix went to a three-day class and learned about problem-solving, time management, and effective communication... but nothing about flying. On the first day of "flying lessons," Clarence could barely control his excitement (and Felix could barely control his bladder). Clarence explained that their apartment had 15 floors, and each day Felix would jump out of a window, starting with the first floor and eventually getting to the top floor. After each jump, Felix would analyze how well he flew, isolate the most effective flying techniques, and implement the improved process for the next flight. By the time they reached the top floor, Felix would surely be able to fly.

Felix pleaded for his life, but it fell on deaf ears. "He just doesn't understand how important this is..." thought Clarence, "but I won't let nay-sayers get in my way." With that, Clarence opened the window and threw Felix out. Felix landed with a thud.

The next day, poised for his second flying lesson, Felix again begged not to be thrown out of the window. Clarence opened his *Pocket Guide to Managing More Effectively* and showed Felix the part about how one must always expect resistance when implementing new programs. With that, he again threw Felix out of the window. (THUD.)

On the third day (on the third floor) Felix tried a different ploy. Stalling, he asked for a delay in the "project" until better weather would make flying conditions more favorable. Clarence, however, was ready for him: He produced a timeline, pointed to the third milestone, and asked, "You don't want to slip the schedule, do you?" From his training, Felix knew that not jumping today would mean that he would have to jump TWICE tomorrow... so he just said: "Okay. Let's go." Out the window he went.

Now this is not to say that Felix was not trying his best. On the fifth day, he flapped his feet madly in a vain attempt to fly. On the sixth day, he tied a small red cape around his neck and tried to think "Superman" thoughts. But try as he might, he could not fly.

By the seventh day, Felix (accepting his fate) no longer begged for mercy... he simply looked at Clarence and said: "You know you're killing me, don't you?" Clarence pointed out that Felix's performance so far had been less than exemplary—

failing to meet any of the milestone goals he had set for him.

Felix said quietly: "Shut up and open the window," and he leaped out, taking careful aim on the large jagged rock by the corner of the building. And Felix went to that great lily pad in the sky.

Clarence was extremely upset, as his project had failed to meet a single goal that he set out to accomplish. Felix had not only failed to fly, he did not even learn how to steer his flight as he fell like a sack of cement... nor did he improve his productivity when Clarence had told him to "fall smarter, not harder." The only thing left for Clarence to do was to analyze the process and try to determine where it had gone wrong.

After much thought, Clarence smiled and said: "Next time... I'm getting a smarter frog!"

That scenario sure fits with evolutionary expectations: that simply by wishin' and hopin' and with enough time, information to form new organs and abilities can be acquired by living creatures out of thin air. Of course this scenario does not include "unlimited" time, but it makes no difference, for evolution's expectations can never be met. Becoming more knowledgeable and informed in evolutionary dogma by spending time in research and studying worthless information will not advance their cause, for no matter how they manipulate the data, rearrange the evidence, or change the time sequence; reindeer can never fly, nor even get off the ground.

GREAT WHITE SHARKS

The *religion of evolutionism* is a caustic and corrosive ideology. For those entrenched in its doctrines, it affects their logic and reasoning abilities. Its poisonous and astringent ideology pushes out *truth, facts,* and *information* conducive to a proper understanding of science. Thus, when evolutionists give instruction in evolutionary concepts and ideas in the classroom or in any media, they are in truth, chroniclers and tellers of tall tales. If you do not think this is so, ponder this.

I was switching channels and happened to catch the last fifteen minutes or so of a nature program about great white sharks. On a table was the vanquished predator, the head of a six-foot-long juvenile white was being cut open to study its brain, which, contrary to common misconception, is not the size of a walnut. From its two huge Y-shaped, bulb-like olfactory organs to its brainstem, the complete shark brain is actually about 2 ft long. Totally unlike a human brain, it is composed of five distinct sack-like compartments that are joined together in almost consecutive fashion.

The Great White shark is a top predator that is uniquely adapted to its ocean environment. Humans have five senses, but sharks have eight highly developed and refined senses: smell, taste, eyesight, touch, hearing, lateral line (sense vibrations in the water), pit organs (not fully understood yet), and what would be a prized power for a superhero, electromagnetism (ampullae of Lorenzini), a sense that can detect an electrical field (such as the electricity given off by a beating heart) that is used to give the exact location of a possible meal. This sense (tiny receptor holes located on a shark's snout) may also serve to detect magnetic fields, which some sharks may use in navigation. Along with these finely honed senses, it sports a sleek, torpedo-shaped body capable of speeds up to 35 miles per hour (50 kilometers per hour). Along with its 300 sharp serrated teeth, arranged in seven rows, this makes it a formidable hunter and killing mechanism. It is no longer considered just a mindless eating machine; many scientists now believe that great white sharks are intelligent and highly inquisitive creatures.

As the scientists analyzed the brain, its parts and structure, they were amazed and awed by the sophistication and complexity of this supposedly ancient organ, which controlled this finely tuned, impressive, and remarkable fish. The narrator called the great white's brain a *super-computer*. He discussed how little scientists actually understood its workings, how advanced and unknown were its mechanisms, and how little we really understood of its systems. Without reason, or direction, or purpose (rational beings have purpose, but not random chance), evolution produced a super-computer, the brain of a Great White shark, which resides in an impressively engineered and complexly designed machine which constructed itself?

Unfortunately for much of the populace, a lifetime of evolutionary brainwashing has caused many who watch such programing to unthinkingly just nod their heads in agreement when this kind of piffle is expounded. Does anyone really believe there is even a slight possibility for a man-made supercomputer to build itself? That if all the raw ingredients needed to build such a machine were placed in the same vicinity and location, or maybe within the confines of the same room, that given enough time, with access to a power source, such as a working electrical outlet, sooner or later when a scientist opened the door to this room, an assembled, working computer would be writing its own programs, doing its own updates and constructing a female machine, so it could reproduce? No, but even if it was able to somehow acquire DNA it could still not reproduce, for at least one other thing is necessary, *life*, one of the components that can never happen by chance or accident.

If this cannot happen, how much more unbelievable is it to think that a living super computer could design, find the raw materials, assemble the parts, and place itself in control of a masterfully engineered living machine that also invented, fabricated, and assembled itself. Truly, this kind of evolutionary logic and reasoning is mystifying.

A FLOOD OF MISINFORMATION

We will continue this chapter with an example of unbridled malice. *Answers in Genesis*, a creation ministry located in the state of Kentucky, built the largest wood-framed structure in the world, an impressive life-size replica of the ark. They call this exhibit the Ark Encounter.

Josh Rosenau, an evolutionary biologist came for a tour, and afterwards wrote a ridiculous, irrational, and nonsensical story about his Ark Encounter. This man is a member of academia and an influential figure in university and scientific circles. He sits high in the councils, fraternities, and brotherhoods of evolutionism. He is one of its High Priests, the Programs and Policy Director of the U.S. National Center for Science Education. Therefore, such a man should clearly understand science, his profession, vocation, and craft. Thus, if he wrote or uttered things contrary to known scientific laws and evidence, would not be doing so ignorantly.

His story of August 5, 2016, "School Field Trips to Creationist Ark? Sink that Idea Right Now," is a deceitful and mendacious attack upon a perceived rival religion. Here is a portion of his diatribe:

FLOOD OF MISINFORMATION

…everything in the park is designed to promote scientifically impossible ideas that contradict everything that scientists know. From astrophysics to zoo-keeping, the visitor is deluged with misinformation. It may be impossible to find a single sign in the park that is free of scientific errors.

To give a single example, Ark Encounter is founded on the notion that all the walking and flying animals alive today descend from specimens caged aboard a boat so unwieldy that it surely would have twisted apart in the roiling waters of a biblical flood. It is a notion that founders on the rocks of genetics, biogeography and naval engineering.

Just as pernicious as the scientific errors and the religious proselytising is a subtler form of indoctrination. The relentless message to visitors is that our world is as fallen and wicked as Noah's, and that the destruction of the flood—including the obliteration of all humans other than a virtuous few—was not just acceptable but praiseworthy.

> ...Ark Encounter presents a message as socially divisive as it is scientifically inaccurate, instilling fear, hatred and hopelessness. Those are lessons no school or parent should want their students or children to take on board.[3]

Upon what information and facts is Rosenau basing his opinion that the Ark was unwieldy and would have twisted apart in the biblical flood? How much time has he spent investigating one of the oldest branches of engineering, shipbuilding? Were his assertions based upon scientific research into the meager information concerning the dimensions of the ark as given in the text of Genesis? Were his conclusions based upon years of study into this ancient science? And how much does Rosenau, or any living man, know about the construction methods used by the very intelligent and highly skilled antediluvian shipbuilders?

Laying aside his venom and unrestrained malevolence, which constitutes much of his tirade, we will zero in on this contention "that all the walking and flying animals alive today descend from specimens caged aboard a boat...that founders on the rocks of genetics..." Let's also leave out the religious aspect of this story, faith and belief in the Flood and the Ark, and deal just with science, his specialty—biology.

He is contending and challenging the idea that all the varieties and kinds of animals alive today could not have descended from represented phyla or taxonomic groups of animals starting 4,500 years ago. He believes that that idea is at odds with the science of genetics. We have spent much time in this chapter dealing with that subject, and from what has been observed in nature, and horticulture, have shown that there are always limits of biological diversity in all organisms.

We have also shown and demonstrated, but not as clearly or distinctly, that diversity within kinds of animals can and at times does happen within a relatively short time. Let us therefore provide a few more observations from history of rapid change in the plant kingdom. These rapid changes also occur in the animal kingdom. Some changes were caused, planned, and instigated for man's own purposes, but other changes that produced numerous varieties just happened randomly.

MCINTOSH APPLES

Rave Landscaping & Discount Plant Center website, 621 Valley View Rd, Dallas, PA 18612

My favorite kind of apple is the McIntosh. As far as I am concerned, when freshly picked off the tree, there are few fruits better-tasting or sweeter when ripe than the unique, crisp, juicy, slightly tart, easily bruised, and thin-skinned McIntosh apple. This variety of apple did not exist before 1811. To this day, no one is certain how those apple seedlings came to grow on McIntosh's farm.

John McIntosh, son of Scottish immigrants, bought a farm in Dundela County, Ontario, Canada. While clearing land in an uncultivated region of his farm, he discovered about a dozen saplings that stood out from the rest—apple trees thriving in a region where apples were not known to grow. He transplanted them into a garden next to his log shanty; all but one died. From that one tree, discovered by accident over 200 years ago, every tree growing McIntosh apples has descended, directly from McIntosh's lone surviving sapling. That original tree survived and bore fruit until 1906.

This is not a unique experience or event, for all organisms have this capacity and ability and it happens frequently, creating a different variety of plant, or animal in an instant, as it were. This ability is not accomplished by new genetic information being created or brought into existence, but is simply a reordering, or recombining of information already residing within. Man has also been instrumental in creating thousands of new varietals and strains of plants over the last few millennia. Fruits such as watermelons, peaches, and bananas, and vegetables such as eggplants, carrots, and numerous other kinds of produce, would be unrecognizable to us in their "original" forms. In many cases if you happened to stumble upon a wild version of a plant such as a carrot, you would not even be able to identify it. Take the peach, for instance: It was at one time a small cherry-like fruit, lean-fleshed, but now is pinkish, big, and juicy.

During the course of the last few millennia, people have observed and studied nature, its flora and fauna, and noted small changes and adaptations within many species and types of organisms. As stated, diversity within animal kind can happen suddenly. From numerous observations, study, and practice, it has also been clearly demonstrated by horticultural and agricultural endeavors, that within the time frame of 4,500 years it is quite reasonable to suppose, and is not a problem, for all the varieties of the animal and plant kingdom to come into existence.

What is really "mind-numbing" and so absurd about Rosenau's single example is not his unscientific, irrational, and foolish contention that it is not possible for all the varieties of life we see all around us to spring forth within 4,500 years or so; to be consistent with his evolutionary doctrine, his contention and argument must be, that it is scientific fact that rocks are the ancestors of all living things, *for once upon a time*, billions of years ago, all life sprang forth from inorganic rocky material strewn about this earth's crust!

Someone really needs to explain to Josh Rosenau that reindeer don't fly!

NOTES ON CHAPTER 4

1. Carl Linn (*Carolus Linnaeus*, 1707–1778) was a scientist who classified immense numbers of living organisms. An earnest Creationist, he clearly saw that there were no halfway species. All plant and animal species were definite categories, separate from one another. Variation was possible within a species, and there were many sub-species, but there were no cross-overs from one species to another (R. Milner, *Encyclopaedia of Evolution*, 1990, p. 276).
2. James Allnutt, "New world record for biggest horse," *The Telegraph*, March 31, 2008, http://www.telegraph.co.uk/news/uknews/1583376/New-world-record-for-biggest-horse.html.
3. Josh Rosenau, "School Field Trips to Creationist Ark? Sink that Idea Right Now," *New* Scientist web site, August 5, 2016, https://www.newscientist.com/article/2100109-school-field-trips-to-creationist-ark-sink-that-idea-right-now/, paras. 8–11.

> There is no doubt that natural selection is a mechanism, that it works. It has been repeatedly demonstrated by experiment. There is no doubt at all that it works. But the question of whether it produces new species is quite another matter. No one has ever produced a species by mechanisms of natural selection. No one has ever gotten near it and most of the current argument in neo-Darwinism is about this question: how a species originates and it is there that natural selection seems to be fading out and chance mechanisms of one sort or another are being invoked.[*]

CHAPTER 5

TRANSITIONAL FORMS AND COMMON ANCESTORS

Words spoken by a wise man and written down in an ancient book over three millennia ago ring true today. It was during the trying of Job's faith, his quest for understanding through the midst of pain and suffering that he sought out the true source of all wisdom. Job states:

> Men know how to mine silver and refine gold, to dig iron from the earth and melt copper from stone. Men know how to put light into darkness so that a mine shaft can be sunk into

---

[*] Dr Coloin Patterson, on the subject of 'Cladistics', in an interview on British Broadcasting Corporation (BBC) television, 4 March 1982. Colin Patterson is Senior Palaeontologist at the British Museum of Natural History, London.

the earth...Men know how to obtain food from the surface of the earth...They know how to find sapphires and gold dust...Men know how to tear apart flinty rocks and lay bare precious stones. They dam up streams of water and pan the gold. But though men can do all these things, they don't know where to find wisdom and understanding. They not only don't know how to get it, but, in fact, it is not to be found among the living. "It's not here," the oceans say; and the seas reply, 'Nor is it here.' It cannot be bought for gold or silver...or precious onyx stones or sapphires. Wisdom is far more valuable than gold and glass. It cannot be bought for jewels mounted in fine gold...Then where can we get it? Where can it be found? For it is hid from the eyes of all mankind; even the sharp-eyed birds in the sky cannot discover it...God surely knows where it is to be found...He established it and examined it thoroughly. And this is what he says to all mankind: Look, to fear the Lord is true wisdom; to forsake evil is real understanding. (Job 28:1–28 TLB)

How well-spoken—how profound—are the words of Job. It is true; mankind can do a plethora of diverse things. But as Job states, wisdom cannot be found in water or in anything man has made or can conceive. Likewise, information does not arise out of the dust, spring forth from a mud puddle, nor float down from the clouds.

Leaving the wisdom of God for a short time let us again consider the sagacity-lacking empire and realm of evolutionary theory. It is where the irrational waste countless hours in devotion to ideas that hinder and thwart scientific advancement, because they pursue meritless and fruitless research to answer questions about the origins of life and the universe that already have a solitary answer, but they refuse to accept that certainty. They explore imaginary events, conditions, and activities that never happened. They conduct useless inquiries into things that have no value, except in their own minds. They seek answers and explanations for things that have been conjured and summoned from a mind where fantasy is real, where nightmares are born, and where a moral compass does not exist. It is a domicile where true science has neither foothold, purchase, nor part, for it is pushed aside in favor of superstitions and beliefs that violate, disregard, and flout reason, logic, and true knowledge. Through their clueless endeavors, they may inadvertently learn a few things about forensics, science, and biology, but the answers they are searching for will never be revealed.

The 1997 science fiction horror film *The Relic* was filled with good special effects, a frightening creature, an eerie museum, some bad language, and gore. It was definitely not a family-friendly film. Penelope Ann Miller played an evolutionary biologist (an oxymoronic term if there ever was one) named Dr. Margo Green. During one scene, a young boy asks her what an evolutionary biologist is. She replies: "Someone who is trying to find out where our tails went." This, in reality, is exactly what evolutionists are attempting to understand, but the term "evolutionary biologist" and her reply as to what it is, is akin to asking how Santa's reindeer fly. The assumption is that reindeer fly (evolutionists always assume people once had tails and evolution actually happened) and we must now do research to find out the reason why. Her answer highlights the irrationality of all evolutionary research. It would be analogous to have tens of thousands of scientists and researchers world-wide over the course of decades in an ongoing effort—doing countless hours of study—to investigate the cause and the reason why Santa's reindeer have the ability to defy gravity and travel through the air at amazing speeds. There is no physical reason, no basis in science or physics that would give cause for a reindeer to fly. The only time reindeer flight might be possible is if one jumped off a cliff. It would then fly all the way down to the ground, where sudden deceleration forces would curtail its flight path and leave it much flatter and a great deal broader than its preflight condition. Of course, that would not be true flight, but what we call free-fall.

Reindeer don't fly; therefore it would be lunacy and a sign of mental illness to pretend that they do and then, based on that evidence, do scientific research to find out how they fly. Evolution did not happen; therefore it is insanity to try and find out how it happened!

MUSH ON, SANTA CLAUS

Now Santa Claus, the musher of Dasher, Dancer, Prancer, and the rest of the reindeer gang, does have an historical basis in fact. At the root of the story stands a true Christian named Saint Nicholas. He was born during the third century in the village of Patara, which is on the southern coast of Turkey. His wealthy parents raised him to be a devout Christian. After the death of his parents, he used his inheritance to assist the sick, needy, and suffering. Down through time, his deeds of mercy, compassion, generosity, and love have fostered countless stories and legends. He was a man of extraordinary character and was greatly revered and beloved as a helper and protector of those in need.

It would take many pages to trace history's long journey from the Fourth Century Bishop of Myra, St. Nicholas—the man whose generosity, kindness and devotion to God knew no bounds—to the jolly, roly-poly, red-suited American symbol for Christmas festivities and commercial activities we now know as Santa Claus. Here are a few of the more notable highlights: Throughout history, and quite frequently during the 1800s, there were changes in his appearance and the saint's name shifted to Santa Claus—a natural phonetic alteration from the German Sankt Niklaus. In 1823, the jolly elf image received a big boost from a poem attributed to Clement Clarke Moore that was destined to become immensely popular, "A Visit from St. Nicholas," now known as "Twas the Night before Christmas." The poem has had enormous influence on the commercialization and Americanization of St. Nicholas. Almost single-handedly, Clement Moore defined our now timeless image of Santa Claus.

Santa Claus has evolved from fact to fiction, whereas evolutionism has never had science, reason, or logic on its side. The core and heart of evolutionism is based on a misunderstanding of science and God's creation. Mix that with delusions of evil men, stir in a little nonsense and folk tales, and it spells out man's rebellion against his Creator. Below is a somewhat lengthy but interesting anonymous ode I came across years ago. It struck me as humorous and will impart some lightness into our conversation:

SANTA CLAUS: AN ENGINEER'S PERSPECTIVE

There are approximately two billion children (persons under 18) in the world. However, since Santa does not visit children of Muslim, Hindu, Jewish or Buddhist religions, this reduces the workload for Christmas night to 15% of the total, or 378 million (according to the Population Reference Bureau). At an average rate of 3.5 children per household, that comes to 108 million homes, presuming that there is at least one good child in each.

Santa has about 31 hours of Christmas to work with, thanks to the different time zones and the rotation of the earth, assuming he travels east to west (which seems logical). This works out to 967.7 visits per second. This is to say that for each Christian household with a good child, Santa has around 1/1000th of a second to park the sleigh, hop out, jump down the chimney, fill the stockings, distribute the remaining presents under the tree, eat whatever snacks have been left for him, get back up the chimney, jump into the sleigh, and get on to the next house.

Assuming that each of these 108 million stops is evenly distributed around the earth (which, of course, we know to be false, but will accept for the purpose of the calculations), we are now talking about 0.78 miles (1.3 km) per household; a total trip of 75.5 million miles (125.83 million km), not counting bathroom stops or breaks. This means Santa's sleigh is moving at 650 miles per second (1083 km/s), 3000 times the speed of sound. For purposes of comparison, the fastest manmade vehicle, the Ulysses space probe, moves at a poky 27.4 miles per second (45.7 km/s), and a conventional reindeer can run (at best) 15 miles per hour (25 km/h)—four thousands of a mile (4/1000) per second (6.9 m/s).

The payload of the sleigh adds another interesting element. Assuming that each child gets nothing more than a medium-size Lego set (2 lbs, or 0.906 kg, that is), the sleigh is carrying over 500 thousand tons US (508,000 t metric), not counting Santa himself. On land, a conventional reindeer can pull no more than 300 lbs (136 kg). Even granting that the "flying" reindeer could pull ten times the normal amount, the job can't be done with only eight or even nine of them—Santa would need 360,000 of them. This increases the payload, not counting the weight of the sleigh—another 54,000 tons (54,864 t metric), or roughly seven times the weight of the Queen Elizabeth (the ship, not the monarch).

600,000 tons (606,600 t metric) travelling at 650 miles per second (1083 km/s) creates enormous air resistance, and this would heat up the reindeer in the same fashion as a spacecraft re-entering the earth's atmosphere. The lead pair of reindeer would each absorb 14.3 quintillion Joules of energy per second. In short, they would burst in flames almost instantaneously, exposing the reindeer behind them and creating deafening sonic booms in their wake. The entire reindeer team would be vaporized within 4.26 thousands of a second (0.00426 s), or right about the time Santa reached the fifth house on his trip.

Not that it matters, however, since Santa, as a result of accelerating from dead stop to 650 miles per second (1083 km/s) in 0.001 seconds, would be subjected to top acceleration forces of 17,500 g's. A 250 pound (113 kg) Santa (which seems ludicrously slim) would be pinned to the back of the sleigh by 4,315,015 pounds of force (195,470 kg force, or 1.9547 MN), instantly crushing his bones and organs and reducing him to a quivering blob of pink goo.

Therefore, if Santa did exist, he's dead now.

After you have mused over and have thoughtfully and carefully considered the logic and arguments used against the "theory" of evolution in this book, you will realize that "theory" is also dead.

UNQUESTIONED BELIEF

Truly, the belief in evolutionism has caused seemingly lucid and intelligent men and women to call fantasies reality and science. What we now have are modern pagans parading their ignorance and fallacious beliefs before the public as truth and science. Through their willing accomplices, book and magazine publishers and television and radio networks, they wield power not much different than the witch doctor or shaman who holds sway over the masses, savage tribes, and ignorant villagers. The evolutionist will always grope in the dark—searching for answers to questions that can never be answered. As Job stated, they have neither wisdom nor knowledge. They have faith in the god titled *Nothing*. The things they see and observe will always be interpreted incorrectly, for they see everything within the context of evolution. It is like looking through sunglasses that are very dirty and scratched; no matter where they look and what they view, it will always look dark, hazy, and somewhat blurry. Without the true source of wisdom, they search but are never able to come to the knowledge of the truth. God is the source of truth, the fountain of true wisdom, and the origin of all true science and knowledge.

Evolutionists know evolution happened; for them, it is an unquestioned fact, reality, and truth, which is why it is always stated with assurance and confidence. The proof of evolution is always the assumption of evolution. They never question whether or not it actually happened. The only question ever entertained is *how* did it occur?

On May 21, 2013, on the website *Random Rationality*, the author Fouret Janabi interviewed Jerry Coyne; the author of *Why Evolution is True* and a professor in the Department of Ecology and Evolution at the University of Chicago. Professor Coyne stated:

> The part that everyone agrees on, let me underline in the beginning, is that *evolution happened*, it took billions of years, *the Earth is 4.6 billion years old*, and life has been here for at least 3.5 billion; that *there is common ancestry of all forms of life because there is a branching bush of life*, and that, in terms of the adaptive character of life was produced by the process of natural selection. So, those are the bedrock foundational principles of modern evolutionary theory, and those have not been called into question... So, in evolution, as in all sciences, there actually are debates between scientists on the details... What are the parts of evolution that are being debated between scientists; *not that as evolution occurred, but how it occurred*. (emphasis added)[1]

The National Academy of Sciences (NAS) 1998 guidebook states the following concerning evolution:

> But how can you be sure that evolution is all that important? Aren't there a lot of scientists who don't believe in evolution? Say it's too improbable? "*The debate in science is over some of the details of how evolution occurred, not whether evolution happened or not.* A lot of science and science education organizations have made statements about why it is important to teach evolution. (emphasis added)[2]

There are basically two arguments that make or break the case for evolution. Every other proof or evidence hinges upon the foundation—

the truth of these arguments. If these two things are neither fact nor proven, then there is nothing of substance for the evolutionist to argue, and if there is no evidence, no facts or observations behind their statements, it is a fairy tale—a belief without any science whatsoever.

Two of the bulwarks of science that stand unquestionably against evolutionism are spontaneous generation, also termed abiogenesis, and transitional forms. This is at the heart of the matter and debate. How these scientists conduct their study and research on bones and fossils would make sense only if evolution was true. You would expect to find and see the gradual change in the structures of animals over time. As I wrote in my book, *Musings on Creation and Evolution*:

> It is always assumed by evolutionists that one kind of creature whose bones are now fossils evolved into another kind, and that you can trace their evolution through the fossil record. That is an assumption that cannot be known with any kind of certainty. The so-called line of decent or lineage, of the progression of one kind of creature into another, is always assumed and never questioned with any real effort. In every interview, every story, article or book dealing with evolution, this unprovable, unknowable "fact" is always assumed and counted as truth, and believed and accepted just as sure as the sun will rise tomorrow morning. It is then on this "fact" that the proof rests, that everything has evolved from a common ancestor. That assumption is a house of cards and lacks any credibility. Neither proof nor any evidence is ever offered.[3]

Every week on the Internet you can read the latest stories the evolutionary community promotes concerning the tale of evolution. This beauty of a story by Laura Gegget, titled "Ancient Knife-Toothed Reptile Is Crocodile Cousin," was posted on January 21, 2015. As we read through her "fact-filled" article we learn the following:

> The fossil of a prehistoric 9-foot-long carnivorous reptile ... is helping researchers fill out the early branches of the reptile family tree. [...] It's unclear where the reptile ... falls on the evolutionary tree.... But the new findings show that it is either the closest relative of the common ancestor of birds and crocodylians ... [...] There's such a huge gap in our understanding around the time when the common ancestor of birds and crocodilians was alive.[4]

Also in January of 2015, Agata Blaszczak-Boxe posted this "informative" story, titled "Oldest Known Snake Fossils Identified":

> New fossils that may be the oldest known remains of snakes are helping researchers to better understand how the slithering beasts evolved ... [...] The new fossils challenge previous theories that suggested that the long, thin skull structure that is characteristic of snakes is a feature that evolved after the animals became legless and developed their elongated bodies ... the new study suggests that "the skull evolved first, and the legless thing followed," said study author Michael Caldwell, a professor ... [...] The remains seem to be most similar to modern snakes that belong to rather obscure groups, and are thought to be somewhat primitive members of the snake family ... [...] "it means that the group had evolved and radiated long before that," [Caldwell] said.[5]

A term used in many of these articles is *common ancestor*. Another favorite phrase is "this find helps us fill in some gaps in the [fill in the type of animal here] family tree." Of course, if they traced the family tree back far enough, the common ancestor would be a rock—which was birthed by the Big Bang, which was brought into existence by its creator, called *Nothing*. The respective arguments bantered back and forth between evolutionists and creationists at times remind me of grade-schoolers arguing over something:

Creationists: "No, there are not transitional fossils!"

> There are no transitional fossils. Evolution predicts a continuum between each fossil organism and its ancestors. Instead, we see systematic gaps in the fossil record.[6]

Evolutionists: "Yes, there are transitional fossils!"

On November 15, 2007, Donald Prothero, a Columbia University lecturer, wrote:

> There are no transitional fossils. Not true—in my new book, *Evolution: What the Fossils Say and Why It Matters*, I document dozens of transitional sequences of fossils, showing the evolution not only of individual lineages, but also of transitional forms that span the gaps between major groups—the "macroevolution" that creationists deny.[7]

> A transitional form is an organism that has features intermediate of its ancestors and progeny. The term is most common in evolution to refer to organisms that show certain features (wings, feathers, gills and so on) partly in development. In theory, every fossil is a transitional form if it has descendants and each living creature is a transition between its parent and its offspring.[8]

> One frequently cited "hole" in the theory: Creationists claim there are no transitional fossils, aka missing links. Biologists and paleontologists, among others, know this claim is false. [...] At least hundreds, possibly thousands, of transitional fossils have been found so far by researchers. The exact count is unclear because some lineages of organisms are continuously evolving.[9]

J: The other evidence is some of the fossil record, the finding of the intermediate whales. When I was in grad school, we knew that reptiles had ancestors to mammals... now we have an even better fossil record because we know that birds evolved from dinosaurs... [...] Same thing with whales, we see this whole intermediate group of whales about 45 million years ago, we have their ancestors and this whole series of animals losing their hind limbs, having their nostrils moved on top of the head, developing flippers, losing their ears, and not only do we have the fossil sequence, but it occurs in exactly the right time. ...the fossil record is clear... [...]

F: Before I read your book, it had been some time since school and evolution in science class. I kept hearing this claim, there are no transitional fossils, the missing links are not there. While I still of course believed and understood evolution, as soon as I read your book, I realized that the fossils actually are there, it's not so much [that they are] not dealing with them, they're just denying that they exist in the first place... It's just fact-denying.

J: Yes, creationists tend to not listen, because if they would listen, they'd give up creationism and become evolutionary biologists, so they maintain creationism, and remember, this is all religiously based.[10]

Evolutionists: "Yes, there are transitional fossils!"
Creationists: "No, there are not transitional fossils!"
Evolutionists: "YES, THERE ARE!!!"
Creationists: "NO, THERE ARE NOT!!!"

An evolutionist who reviewed this book shared these thoughts about the previous discussion. She did not think the authors of these quotes were arguing at cross purposes, but the general understanding of the subject was becoming clearer as the technology to examine them improves. Furthermore, the same process that improves the development of chemistry and medicine, in disciplines such as oncology and pharmacology, helps us to better treat specific illnesses and improves our understanding of their natures. Likewise, these same developments in understanding are echoed in the study of evolution, for the more we know how evolution occurred, we are better able to see and piece together the fossil record based on the number and the chronological ages of the fossils recovered. However, she fails to see that at the very foundation of her rebuttal is the assumption that evolution happened, and thus there is a chronological order and evidence to study and learn from.

This debate between these two religions, which we have touched upon briefly, is out in the open and very public. There is, however, a side of this dispute that the evolutionists would like to keep hidden and undisclosed to the public. Sometimes in moments of candor, within the secretive conclaves of those high in the councils of comradeship, pejorative information will come to light that is not customarily disseminated among the general population.

Author Luther D. Sunderland happened across such evidence when he was writing his book, *Darwin's Enigma: Fossils and Other Problems*. During his research, before interviewing Dr. Patterson, he had read his book *Evolution*, which he had written for the British Museum of Natural History. In it Patterson had solicited comments from readers about the book's contents. In his own book, Sunderland included Patterson's reply to one reader who asked why he did not put a single photograph of a transitional fossil in his book. On April 10, 1979, he replied to that reader in a most candid letter, as follows:

> I fully agree with your comments on the lack of direct illustration of evolutionary transitions in my book. If I knew of any, fossil or living, I would certainly have included them. You suggest that an artist should be used to visualize such

transformations, but where would he get the information from? I could not, honestly, provide it, and if I were to leave it to artistic license, would that not mislead the reader?

I wrote the text of my book four years ago. If I were to write it now, I think the book would be rather different. Gradualism is a concept I believe in, not just because of Darwin's authority, but because my understanding of genetics seems to demand it. Yet Gould and the American Museum people are hard to contradict when they say there are no transitional fossils. As a palaeontologist myself, I am much occupied with the philosophical problems of identifying ancestral forms in the fossil record. You say that I should at least "show a photo of the fossil from which each type of organism was derived."? *I will lay it on the line—there is not one such fossil for which one could make a watertight argument.* The reason is that statements about ancestry and descent are not applicable in the fossil record. Is Archaeopteryx the ancestor of all birds? Perhaps yes, perhaps no there is no way of answering the question. It is easy enough to make up stories of how one form gave rise to another, and to find reasons why the stages should be favoured by natural selection. But such stories are not part of science, for there is no way of putting them to the test. (emphasis added)[11]

It is unfortunate that this kind of information, which truly reflects what is found concerning the fossil record, is not better known. Some contend that Patterson's comments are over thirty years old and therefore not valid, for new discoveries and information has come to light. Water freezes at 32°. That fact was true 4,000 years ago, and will be true throughout all eternity. Likewise, though Patterson's letter was written 30 years ago, since evolution cannot occur, it will remain a scientific fact throughout the coming ages, and no new information can ever change that. Of course, his statement contradicts Coynes' pronouncement, but remember, evolutionists only believe and have faith in transitional fossils; they are not relying on unbiased evidence. As to why he would acknowledge such a thing, his comment is really not that mystifying. Consider this: Everyone in their lifetime has spoken words they would like to retract, written words they would like to delete, and for Patterson, I am sure he would have liked to expunge that incriminating letter.

One evolutionist, in an attempt to refute these comments, introduced me to the work of Neil Shubin, a prominent University of Chicago paleontologist and Professor of Anatomy, whose claim to fame is the discovery of *Tiktaalik roseae*, a creature that supposedly represents the evolutionary transition from fish to amphibians and, much later on, your mother-in-law. He has put together a three-part television series, *Your Inner Fish*, which you can watch on YouTube. Episode 2 starts out this way:

> From the plains of South Africa, to the shores of Nova Scotia, in the bones of ancient creatures, and deep inside your DNA, lies an incredible story, the story of your body and why you're built the way you are, your skin, your hair, your complex teeth and remarkable sense of hearing, can all be traced back to ancient reptiles that once ruled the earth. Their bodies were shaped by great transitions in the history of life, and that legacy still shapes our bodies today.[12]

Incredible story indeed, copious chance transitions...those ancient reptiles must have had great expertise and training in programing, biology, and other life sciences, for they laid an amazing foundation upon which their progeny would be able to transform and convert into during the coming ages. Yes, what they built was able to be maintained by these wonderful and intelligent lizards, and improved upon during the eons and produce an extraordinary and remarkable world full of wonderful creatures.

The narration continues:

> My name is Neil Shubin, as an anatomist I look at human bodies differently than most people. Within us I see the ghosts of animals past. Distant ancestors who shaped our anatomy in surprising ways.

Shubin is not looking through the clear glasses of science, nor corroborating evidence, but faith in a story. Every bit of evidence found, information learned, and research done by Shubin, can never be objective or non-biased, for evolutionary lens and blinders are always placed over his eyes before he opens them. He may think he sees something, but that is his indoctrination in evolutionary ideology speaking, in reality all he is seeing is just a mirage, not what really is.

It is true; what evolutionists call transitional fossils are found all over the earth. Evolutionists believe they are transitional, for everything has a common ancestor. Thus every fossil and every bone that is found of some extinct animal can be placed and will fit somewhere upon the phylogenetic tree; as referenced and confirmed by the two stories, the only question asked when a fossil is found is where it falls on the evolutionary tree.

Their faith in evolution is so intimately entwined into the very fiber of their study, that even their questions, research, and musings are tainted and colored by their ideology. It is at this point, stage, and juncture in their argument we must take a stand. Other than their faith and belief in evolutionism, upon what science is that faith confirmed or justified?

It is clearly implied and understood in every story dealing with evolutionism that belief in the "branching bush of life" is built upon faith in evolutionism. One argument confirms the other. But would proof that was built upon such circular reasoning hold up in a court of law? What kind of independent studies or scientific tests have been performed upon fossils that confirm one is the ancestor of another or that all living things have a common ancestor? How, besides their faith in evolutionism, do they determine the pedigree, ancestry, genealogy, and family extraction of a fossil?

In his book *Bones of Contention*, Marvin L. Lubenow points out that:

> The studies that seek to prove that human DNA evolved from chimp DNA start with the assumption that chimp DNA represents the original condition (or close to it) from which human DNA diverged. That is circularity with a vengeance.[13]

In reality, evolutionists have no objective way of testing the genetic relationship, pedigree, or genealogy of fossils, for experimental verification of a fossil's ancestry is not possible. Even if a supposed family of individuals was found, there is no conclusive way to determine their relationship. Genetic testing is a viable science, but to use it to connect so-called families of creatures and their relationships going back thousands or millions of years, as supposed by evolutionists, is overstepping its creditability to a large measure. Lubenow clearly highlights this fact:

> At this point I merely want to emphasize a phenomenon that seems almost universally unrecognized: *Any series of objects created by humans (or God) can be arranged in such a way as to make it look as if they had evolved when in fact they were created independently by an intelligent being.* The fact that objects can be arranged in an "evolutionary" sequence does not prove that they have a relationship or that any of them evolved from any of the others. (emphasis his)[14]

Read these quotes from "Evidence of Common Descent," a Wikipedia article, and see if you can recognize the phenomenon Lubenow stated, which I have numbered for ease of reference:

1. More closely related species have a greater fraction of identical sequence[s] and shared substitutions compared to more distantly related species.

2. The simplest and most powerful evidence is provided by phylogenetic

reconstruction. Such reconstructions, especially when done using slowly evolving protein sequences, are often quite robust and can be used to reconstruct a great deal of the evolutionary history of modern organisms (and even in some instances of the evolutionary history of extinct organisms) . . .

3. While a minority of these elements might later be found to harbor function, in aggregate they demonstrate that identity must be the product of common descent rather than common function.

4. It is possible to find out how a particular group of organisms evolved by arranging its fossil records in a chronological sequence.[15]

In quote number 1, we read about closely and distantly related species. Evolutionary doctrine has a built-in flaw: It must assume that one species is related to another, because there is common descent from a common ancestor. They are working within those parameters, assumptions, and considerations, so any conclusions or "evidence" must always fit within that framework.

Quote number 2 again works within the belief that there is an evolutionary history of life, so any evidence that does not fit within that premise never will nor ever can be considered. They are conforming and adapting all of the evidence to their belief system.

In number 3, the belief in evolutionism colors their conclusion that it must be the product of common descent rather than common function. Have they even considered that the evidence might point to a common designer?

In reference to the last quote, how would they know the chronological sequence of an organism? Sometimes the strata in which the fossils are found are in an inverted order (according to evolution) and are not found in proper sequence.

> A fossil cannot be dated by itself; it can only be dated by knowing where, in the geologic column, it was found. A layer of rock cannot be dated without knowing what fossils are found in it. This is why we cannot accurately date anything

using the geologic column. "*The rocks do date the fossils, but the fossils date the rocks more accurately.* Stratigraphy cannot avoid this kind of reasoning if it insists on using only temporal concepts, because circularity is inherent in the derivation of time scales." (emphasis added)[16]

However, no matter where or in which order fossils are found, scientists arrange the fossils according to the evolutionary order. They do not arrange the fossils based upon any evidence found in the fossils themselves; they arrange them by their belief that the fossils have a chronological sequence. Even the assumption that there is a chronological sequence is begging the question, for any so-called chronological sequence would be based upon the belief that all things are related and have a common ancestor.

VESTIGIAL ORGANS

Many evolutionists have earned PhDs in their respective fields of expertise. However, what advantage do those hard-earned degrees give them in the real world (we are not referring to those fields involved in engineering, medical techniques, genetic testing etc., where the scientific method can be used), out in the *field*, where much of what they "learned" provides no value, has no basis in fact, does not fit with what is observed in the natural world, and, in many cases is contrary to the laws of science? Of course, at times those who believe in evolution do make contributions to science, but that would be despite the hindrance of evolutionary doctrine.

Evolutionary presuppositions contaminate, taint, and flaw all academic studies and research done by evolutionists. A perfect example (out of many) of this is the belief in so-called vestigial organs. Most secular biology textbooks will include a section on vestigial structures, supposed remnants left over from our ancestors that have lost their original function or capacity. This idea has impeded medical research for years.

The appendix has long been thought of as a useless dead-end organ—an idea and ignorance left over from Darwin's time. However, just because you can remove it and live without it, does not mean it was useless; people have had limbs amputated and still remain useful, productive, and happy members of the human race. Many of the functions and purposes of that wonderfully designed organ are now beginning to come to light. Among other functions, it helps keep the body healthy by manufacturing antibodies. It is also rich in lymphoid tissue, which acts as a filter to remove bacteria and helps protect the intestines against infection.

NEANDERTHAL AND CRO-MAGNON MAN PRECONCEPTIONS

In the Wikipedia article "Scientific Evidence," in the section on principles of inference, we read:

> A person's assumptions or beliefs about the relationship between observations and a hypothesis will affect whether that person takes the observations as evidence. These assumptions or beliefs will also affect how a person utilizes the observations as evidence.[17]

A perfect case in point of how evolutionary beliefs have affected how researchers utilize observations is seen in the analysis of man's supposed evolutionary predecessors, Neanderthal and Cro-Magnon man. Modern prognosticators of evolutionary religion proclaim that these two groups of peoples were less than human. Please read a sample of their assertions:

> The scientific name for modern humans walking the Earth today is *Homo sapiens sapiens*. Most scientists today describe Neanderthals, which are extinct, as *Homo neanderthalensis*—a species distinct from modern humans —though some scientists still hold that Neanderthals were a subspecies of *Homo sapiens*. They looked pretty similar to modern humans,

but were stockier, with more prominent, sloping brows and wide noses.

Neanderthals are also our closest evolutionary relative, and scientists know that at some point tens of thousands of years ago, Neanderthals and Homo sapiens mated to some degree.[18]

Here is another example of what fully biased religious individuals believe about our ancestors:

> Imagine trying to describe an entire culture based on a few skeletons. Now imagine that those bones are 40,000 years old. [...] But scientists in Spain and Germany are starting to do just that. These scientists are studying *Neanderthals, a group of human-like creatures* that lived in Europe and western Asia starting half-a-million years ago. They overlapped for a bit with the ancestors of modern-day humans. (emphasis added)[19]

Recent research into Neanderthals has determined, while comparable in height to modern humans, they had a more robust build, were much stronger, with particularly strong arms and hands. According to new research, their eyesight may also have been better, owing to larger eye sockets and larger areas of brain devoted to vision. These red- and blond-haired, light-skin-toned peoples (as new studies in genetics have concluded), also had a large cranial capacity, which at 1,600 cm is notably larger than the 1,400 cm average for modern humans, indicating that their brain size was larger.[20]

Another supposed early human is Cro-Magnon, a common name that has been used to describe the first modern humans that lived in Europe. Like Neanderthals, they were also robust and powerful, for they had very strong, muscular bodies. Their brain capacity was also larger, about 1,600 cm, which is about 98 cu larger than the average for modern humans.[21]

What modern body builders wouldn't give to have Cro-Magnon or Neanderthal genes, for what a powerful muscular physique and form it would create after training. If scientists could clone a Neanderthal, and Arnold would be there to help "pump him up," the annals of body-building and weight-lifting would be forever changed. Records would fall, statistics would need to be updated, for the world's best would now far exceed by a wide margin any previous achievements.

Our supposedly less evolved early ancestors had a body that was physically superior to modern man, so why not an advanced mind to go along with it? Of course, a larger brain does not necessarily translate to more intelligence—or does it? An article in Wikipedia titled "Brain Size" relates the following:

> Studies demonstrate a correlation between brain size and intelligence, with larger brains predicting higher intelligence. It is, however, not clear if the correlation is causal. The majority of MRI studies report moderate correlations around 0.3 to 0.4 between brain volume and intelligence. The most consistent associations are observed within the frontal, temporal, and parietal lobes, the hippocampus, and the cerebellum, but only account for a relatively small amount of variance in IQ, which suggests that while brain size may be related to human intelligence, other factors also play a role. In addition, brain volumes do not correlate strongly with other and more specific cognitive measures.[22]

Another article, also from Wikipedia, entitled "Neuroscience and Intelligence," states:

> Overall, larger brain size and volume [are] associated with better cognitive functioning and higher intelligence. The correlations range from 0.0 to as high as 0.6, and are predominantly positive. The specific regions that show the most robust correlation between volume and intelligence are the frontal, temporal and parietal lobes of the brain. Therefore it can be safely concluded that larger brains predict greater intelligence.[23]

After reading this information, it does seem reasonable to assume those so-called early hominids, those supposedly less evolved beings, were after all modern man's equal or maybe even his superior when it comes to intelligence.

When reading through the information concerning these two ancient groups of peoples, it seems the only thing primitive about them was their lifestyle. This reminds me of another group of people, contemporary with modern man, the indigenous peoples who inhabit Australia, the Aborigines. Many Aborigines still live a somewhat primitive lifestyle. They simply choose to do so. They can survive in a harsh land, the Outback, for they understand and know their way around the Bush. They can also survive within the modern world and fit right in with the rest of us. They are not sub-human, but are as intelligent as any other group of peoples found throughout the Earth. And that cloned Neanderthal, after a shower, close shave, styled-haircut, a few dashes of Caron's Poivre, and spending a good chunk of change on an Ermenegildo Zegna, an Italian hand-made, hand-stitched, and tailored business suit, he would look like the CEO of a Fortune 500 company, maybe your favorite college professor, or the beefed-up twin brother of bodybuilder Lou Ferrigno. According to the Wikipedia article "Brain Size":

> From early primates to hominids and finally to Homo sapiens, the human brain has continued to grow. The volume of the human brain has increased as humans have evolved ... starting from about 600 cm^3 in Homo habilis up to 1,600 cm^3 in *Homo neanderthalensis, which was the hominid with the biggest brain size. The increase in brain size stopped with neanderthals. Since then, the average brain size has been shrinking* over the past 28,000 years. The cranial capacity has decreased from around 1,550 cm^3 to around 1,440 cm^3 in males while the female cranial capacity has shrunk from around 1,500 cm^3 to around 1,240 cm^3. (emphasis added)[24]

So, according to evolutionists, early primates evolved into early hominids such as Cro-Magnons and Neanderthals and finally to us, Homo sapien sapien, the top of the evolutionary ladder, the pinnacle of evolutionary success. But does the information given fit this evolutionary story? Do the facts and evidence related so far line up with their narrative? Those early hominids were subhuman, less evolved than modern man, more primitive, simpler in design, for they were not as far along on the ladder of evolution? Did I miss something in these articles? Does it not seem to you that modern man's intellect has been degenerating and not improving, for his brain size has been shrinking? It is true, the knowledge man has learned and has accumulated down through the years has grown, but man's intellect, his mental power, has certainly taken a big hit—and physically, we are no match for the average Neanderthal or Cro-Magnon.

I suppose, with evolutionary logic, you can prove anything. Man's brain was growing larger over time, which meant he was evolving into a more sophisticated, advanced, and higher kind of being. Man's brain is now growing smaller over time, which means he is evolving into a more sophisticated, advanced, and higher kind of being. Evolutionists must be in a real stupor, for they do not yet realize that their goose has now been fully cooked; in fact, it's in flames... and about to burn their house down. From the evidence, it seems the zenith, the apex of the evolution of the human species took place and culminated with those early hominids, and everything has been down-hill since then.

History, facts, and evidence more than suggest that the Scriptural account of early man, pre- and post-Flood, is the correct and true narrative. Those bones and artifacts found in 1856, in a cave in the Neander River Valley of Germany, of people now known as Neanderthals, were the descendants of Noah, probably a few generations removed from that patriarch.

The Neanderthals stepped onto, and ventured out into an environment vastly different than the world Noah departed from when he entered the Ark. The lushly forested landscapes, abundant food supplies, spring-like climate over much of the earth, and bountiful flocks and herds of wild creatures were gone. It was a harsh world inherited by the Neanderthal clan. Unlike before, there were now drastic fluctuations in temperatures. In some places there was unbearable heat, while in others extreme cold. Snow fell for the first time, and started to accumulate on some of the newly formed mountains, giving rise to glaciers. Game was scarce, and other food stuffs were also in short supply.

Did the Neanderthal tribe just happen to journey and set up housekeeping in a land where even their superior minds and bodies found hard to sustain life? What happened to them? Some of their bones provide evidence of vitamin deficiencies, so did they succumb to that, or some kind of disease? Did they just move on, and are now lost to history? Or maybe a natural disaster or another warring tribe wiped them out. We may never know for sure.

There is, however, one thing about them that is certain: they were 100 percent human, not a sub-species of human. So, despite the contrary evidence, why have the evolutionists come up with this outrageous, nutty, and harebrained idea that they were not fully human? The physical evidence we have discussed sure supports my contention that they were human, and fine specimens at that! What scientific evidence did evolutionists use—what facts, data, or proofs did they rely on for support for their tale?

Dennis O'Neil, in his article "Neandertals," clearly states the reason:

> What finally convinced the scientific community that Neandertals were very ancient Europeans was a combination of additional fossil discoveries and new perspectives that largely began with the publication of Charles Darwin's *On the Origin of Species* in 1859. This seminal work in biology popularized the idea that species of living things evolve over time as a result of natural selection. *Subsequently, it was not a major leap in understanding to realize that humans also must have evolved from earlier forms.* In fact, Darwin proposed just

that in his 1871 influential publication *The Descent of Man, and Selection in Relation to Sex*. For enlightened Victorians, the Neandertals ultimately came to be seen as important human ancestors. (emphasis added)[25]

It was simply the belief (evolutionism) that man had evolved from lower forms of life and not science, scholarship, or facts, which allowed evolutionists to turn modern man's superior into a brutish and bestial simian creature, something which was not yet fully human. This statement does not ignore the decades of research and advances in techniques of chemical and genetic analysis that have occurred since the time of Darwin, for, as I have pointed out, you cannot find evidence that never existed in the first place.

On the *Age of Truth TV*, Lucas Alexander interviewed Dr. Niels Harrit, who is now retired after 42 years as a scientist and Associate Professor Emeritus of Chemistry at the University of Copenhagen. Dr. Harrit stated:

> Natural science is based on experience... There is no truth in science, there's only the best hypothesis of the day. So you have a series of observations and the best model is the truth of the day. You should be ready to take other observations in and eventually correct your hypothesis.[26]

The debate between creationists and evolutionists is not about the fossil evidence. In many instances, the same fossils, remains, data, or bones are used in study and research. After the conclusion of their separate and independent studies, analyses, and investigations, the same fossils might be used by both sides to demonstrate, verify, and substantiate claims. That is why Duane T. Gish could write a book titled *Evolution: The Fossils Say No!* in 1979 and in 2008 Donald Prothero could write a book named *Evolution: The Fossils Say Yes!* Each author looked at similar evidence, but through the lens of a different ideology, dogma, and philosophy. An evolutionist builds his or her facts and evidence upon a belief that operates within a predetermined set of assumptions. Never can an evolutionist question or deviate from these conjectures, suppositions, notions, nor the evolution manifesto. It is also fair to say that scientists who do not share the evolutionists' bias have a different point of reference, build their opinions and reasoning upon another foundation, the underpinning and basis of which is always pushed aside by evolutionists' narrow-mindedness, pride, and rebellion.

Those scientists who hold to a Creator, with a creation that was obviously well designed and organized, wonder greatly when those of Darwin's ilk discuss nature in terms of its superb architecture, structure, design, and engineering but become so irrational, inconsistent, and illogical by attributing what they see and observe to a twist of fate or happenstance. There are a few who believe in a Creator, along with some kind of evolutionary causes which helped to create our present Earth and universe—but that is something akin to trying to homogenize fire and water.

IT'S NOT ROCKET SCIENCE

Werhner von Braun (1912–1977) is probably the greatest rocket scientist in history. He was a German-born American aerospace engineer and space architect and is considered one of the "Fathers of Rocket Science." As the head of NASA's Marshall Space Flight Center, his crowning achievement was to lead the development of the Saturn V booster rocket, which helped land the first men on the moon in July of 1969. In February of 1976, during an interview, Dr. Braun was quoted saying the following:

> After years of probing the spectacular mysteries of the universe, I have been led to a firm belief in the existence of God. The grandeur of the cosmos serves only to confirm my belief in the certainty of a creator. I just cannot envision this whole universe coming into being without something like divine

will. The natural laws of the universe are so precise that we have no difficulty building a space ship to fly to the moon and can time the flight with the precision of a fraction of a second. These laws must have been set by somebody.

Dr. von Braun went on to say:

> "science" and "religion," properly understood, are not antagonistic pursuits. On the contrary, he affirmed, they are "sister disciplines." Through the scientific method, one learns more about "creation," whereas, by virtue of the study of religion, one gains a greater insight into the "creator." By employing the tools of science, man attempts to harness the forces of nature which surround him; through religion, on the other hand, he endeavors to control the "forces of nature," which are at work within him.[27]

It has been said that without men of faith—faith and belief in the Creator of all we see—modern science would never have been birthed. As we look back through history at the pioneers and founding fathers of modern scientific disciplines, numerous are the Christians we uncover.

Leonardo da Vinci (1452–1519) was a sincere believer in Christ and the Scriptures, though he is better known for his incomparable paintings. He was also a great engineer and architect. He is considered by many to be the real founder of modern science. Other Christian fathers of modern science are household names and some are known only within scientific circles—such as Johann Kepler (1571–1630), the founder of physical astronomy. Francis Bacon (1561–1626), the Lord Chancellor of England, is considered by most to be the man primarily responsible for developing the scientific method. Blaise Pascal (1623–1662) is considered the father of the science of hydrostatics and one of the founders of hydrodynamics. On and on the list goes. In Alfred M. Rehwinkel's book *The Flood*, he states:

At a meeting of the British Association of Scientists held in 1865 a manifesto was drawn up and signed by 617 men of science, many of whom were of the highest eminence, in which they declared their belief not only in the truth and authenticity of the Holy Scriptures, but also in the harmony of Scripture with natural science.[28]

Rehwinkel further states that a copy of this manifesto was deposited in the Bodleian Library of Oxford. The text reads as follows:

We, the undersigned students of the Natural Sciences, desire to express our sincere regret that researches into scientific truth are perverted by some in our own times into occasions for casting doubt upon the truth and authenticity of the Holy Scriptures.

We conceive that it is impossible for the Word of God as written in the book of Nature, and God's Word written in Holy Scripture to contradict one another, however much they may appear to differ.

We are not forgetful that physical science is not complete, but is only in a condition of progress, and that at present our finite reason enables us only to see as through a glass darkly, and we confidently believe that a time will come when the two records will be seen to agree in every particular.

We cannot but deplore that Natural Science should be looked upon with suspicion by many who do not make a study of it, merely on account of the unadvised manner in which some are placing it in opposition to Holy Writ.

We believe that it is the duty of every scientific student to investigate Nature simply for the purpose of elucidating truth, and that if he finds that some of his results appear to be in contradiction to the written Word, or rather to his own interpretation of it presumptuously affirm that his own conclusion must be right, and the statements of Scriptures wrong. Rather leave the two side by side until it shall please God to allow us to see the manner in which they may be reconciled; and instead of insisting upon the seeming differences between Science and the Scriptures, it would be as well to rest in faith upon the points in which they agree.[29]

– Cartoon Studios –

Many of these Christians stood tall among their peers during their lifetime. Throughout recorded history, theirs were some of the best and the brightest scientific minds. They were great, for they knew that there was a God who created the principles, the laws, and the order of the universe. Most importantly, they considered themselves privileged, not deserving, to be the ones to help unlock a few of the mysteries of His marvels and wonders. Numerous[30] are the scientists who place their faith in a God of order, reliability, and steadfastness; the religious beliefs of evolutionists, on the other hand, stand in stark contrast to those who were in large part reasonable for the science that has served to benefit all mankind.

It has been said that "beauty is in the eye of the beholder." In the same way, transitional fossils only exist within the mind of the evolutionist. They occur only when you accept the phylogenetic tree. Evolutionists do not see through the eye of evidence; they see through faith in an idea—a belief that causes them to manipulate, conform, adapt, and arrange what they see into a progression of fossil life, from supposedly simple to complex.

Many people assume scientists bring to the table impartiality—a non-biased, non-prejudiced attitude toward their respective studies. However, the opposite is true; before the start of every investigation, all evolutionists begin their studies with inflexible, rigid conclusions and preconceived ideas. This flaw keeps their research subjective and shrouded in bias, for they refuse to consider the possible merits of an alternative point of view. Because of variables not considered in the calculations, or maybe because of the variables themselves, the statistics will always be distorted. That being the case, all so-called evidence proving evolutionism will always lack neutrality, and any research findings will be subjective and one-sided. At times, devious detectives who believe a suspect is guilty, when presented with contradictory evidence, will conceal, manipulate, and hide any fact that supports the innocence of the accused. Likewise, all evolutionists allow their prejudice to get in the way of sagacity, knowledge, and true wisdom.

Let us now tear down another bulwark in the crumbling hovel of evolution: *spontaneous generation*. Of late, the use of that term has become politically incorrect among evolutionists, so we will use the newest contemporary catchphrase, which is now in vogue among the fashionable in-crowd: *abiogenesis*. Evolutionists argue that there is a big difference between these words. An online blog post "Abiogenesis is Not Spontaneous Generation, Period" stated:

> [Abiogenesis] predicts that, as life is made up of chemical reactions, and the constituent components of life can self-arrange given certain conditions, there is some point in Earth's early history wherein a chemical chain reaction went runaway and breached the fuzzy barrier between chemistry and biology. All biology is is [sic] one single long, un-broken chemical reaction that can be traced back to whatever initial condition sparked it billions of years ago.[31]

Evolutionists believe there is absolutely no need for a personal directing intelligence to orchestrate the creation of the vast and majestic grandeur of life. *Abiogenesis* depends upon long, logical processes—a staircase of small steps and natural selection through chemical reactions used to form the great diversity of life we see all around. It was natural selection that led to the survival of the quicker self-replicating, self-arranging, amino acids, and other materials, which then developed into simple bio-machines. These simple machines, after some four billion years, developed into all the forms of complex life that now inhabit this planet.

After the demise of the belief in spontaneous generation in the nineteenth century, it was not until 1924 that the faithful devised another scheme in attempt to keep their beliefs "viable." A Russian biochemist named Aleksandr Ivanovich Oparin presented his ideas in his publication *The Origin of Life* (1953). He taught that life developed very slowly and gradually from chemical evolution of carbon-based molecules within a primordial "broth." Scientists continue to research the supposed specific sequence of chemical events that led to the organization of the first organic molecules, called nucleic acids.

Most evolutionists now deplore the debunked belief in spontaneous generation; many admit it was disproven by Pasteur and others. They also claim that spontaneous generation did not address the actual origins of life—assuming, rather, the generation of evolved complex organisms on a daily basis from non-living material—while the theory of abiogenesis presents a "solid" scientific theory that supposes the slow formation of primitive organisms over hundreds of millions of years.

In actuality, abiogenesis and spontaneous generation are two pieces from the same pie of time. The only real difference between both terms is the elegance of the semantics. Both lexes can be summarized very simply and accurately this way: The emergence of life from non-living, inorganic materials over an indeterminable period of time. So what is the real difference between these two words? Nothing, other than the amount of time it takes the magic of time to change a blob into Bob. Or, to put it another way, it is the length of time it takes for a princess's kiss to change a frog into a handsome and charming young prince.

The last example I will use comes from the online *Encyclopaedia Britannica* article titled "Modern Conceptions of Abiogenesis." While this story is not really any different in tone than most of the narratives put out by the narrow-minded elite crowd, during the writing of this book I had what you might call an epiphany—a sudden intuitive leap into understanding the mindset of the typical evolutionist.

Their writings ooze—nay, radiate—a disdain for, sense of superiority over, abhorrence of, and total disrespect for all things holy. They are intolerant, haughty, contemptuous, and, for the most part unaware that these impulses are reflected in their research. This, in turn, blinds their eyes to their own inconsistencies, biases, and illogical conclusions, for the sieve of evolutionism strains out the seed of truth and keeps it from germinating and producing the fruit of comprehension, understanding, true education, and knowledge. Please read this quote and see if you observe the biases and preconceived ideas and beliefs assumed by its author:

> Research on abiogenesis has benefited significantly from astrobiology, the field of study concerned with the search for extraterrestrial life (life beyond Earth) and with understanding

the conditions required for life to form. Astrobiological investigation of the moon Titan, for example, which has an atmosphere lacking free oxygen, have revealed that complex organic molecules are present there, offering scientists a glimpse into the formation of biological materials in a prebiotic habitat resembling that of early Earth.[32]

Scientists have an understanding of the conditions required for life to form? And upon what observations is this understanding based, and what kinds of tests confirm this knowledge? What kind of high-powered, past-revealing crystal ball have these folks peered into that revealed the conditions on earth billions of years ago? I have got to get me one of those!

Titan's moon atmosphere lacks free oxygen; it is therefore, a glimpse into the conditions of Earth's early habitat. Really? And they know this because...? They "know" it because evolution happened and they know evolution is a fact, for earth had a prebiotic habitat with complex organic molecules; and they know this because that is the way life started on the early Earth; and they know this for sure, because they know evolution happened...

How does this paragraph, with its swirling merry-go-round of circular reasoning (they first assume evolution happened, and work from that premise), help you in your understanding of science, of the workings of the universe and of nature, or in the study of life? How does this nonsense benefit or advance scientific knowledge of the unknown? While there may be a smidgen of science hidden somewhere within this paragraph, there is little substance or beneficial knowledge included within the entire 719 words of the complete article.

Back in 1976, Johnny Cash had a number one hit titled *One Piece at a Time* on the Billboard Hot Country Singles chart. The song is about a man who works on a Cadillac assembly line day after day. Knowing he will never be able to afford one, he decides to steal one by taking one piece at a time. Some small parts were smuggled out in his lunch box and the bigger parts were removed in various other ways. It takes him over 20 years to acquire all the necessary parts. Once he thinks he has a complete set of car parts, he attempts to assemble the pieces, only to find out all those parts from the different model years do not fit together very well, but when adapted and finally assembled, they created a very odd-looking Cadillac. The song closes with him taking his wife for a spin in his Cadillac and everyone "laughin' for blocks around." The last line asks; "What model is it?" to which his answer is, "Well, it's a '49, '50, '51, '52, '53, '54, '55, '56, '57, '58, '59 automobile. It's a '60, '61, '62, '63, '64, '65, '66, '67, '68, '69, '70 automobile." This is kind of like the evolutionists' assembly line, building one simple bio-machine at a time until it is fully adapted. Unfortunately for evolutionists, there is no evidence that life started from simple entities that developed and amalgamated into more complex bio-machines over a long period of time.

A single cell is an irreducibly complex molecular machine with billions of necessary electrical connections. Most cells are composed of these modules: a Golgi apparatus, a cell membrane, ribosomes, lysosomes/vacuoles, an endoplasmic reticulum, mitochondria, primary cilia, a nucleus, and cytoplasm. Could a cell get along without its lysosomes/vacuoles, which break down waste and debris? Is a cell able to function without its ribosomes, the cells' protein-making factories, or the cells membrane or its cytoplasm? [It is possible to remove certain organelles from a cell without damaging its function in any way; however, the cell will itself produce more of the missing organelles and carry on as normal at full capacity.] The cell is a complete system, a single composite unit, in which each part is needed. Unlike that of a Cadillac, which, compared to biological life, is very simple, human "assembly lines" are immensely complicated and require a great deal of planning and oversight. However, no one was in charge of the evolutionist's assembly-line processes.

One evolutionist who read this paragraph questioned my reasoning concerning the cell. She stated:

"Three-parent babies" are the result of replacing the mitochondrial DNA in the mother's egg with the mitochondria from a donor; if a single cell was irreducibly complex, this would not be possible, because removing the mitochondrial DNA from the mother's egg would destroy its ability to function, according to this assertion, but the first such child conceived using this procedure was born in January 2016 and is reported to be healthy.

If I understand the procedure, how does this method rebut my contention? All those scientists have done (a very involved technique), is swap fully functioning parts, the DNA from one egg with the DNA of another. Isn't this analogous to replacing the motherboard in one computer with the motherboard from a different machine?

NOTES ON CHAPTER 5

1. Fouret Janabi, Interview with Jerry Coyne, Random Rationality web site, May 21, 2013, https://randomrationality.com/2013/05/21/why-evolution-is-true-with-jerry-coyne/
2. National Academy of Sciences (NAS), "Chapter One: Dialogue, The Challenge To Teachers," *Teaching about Evolution and the Nature of Science* (Washington, D.C.: National Academy Press, 1998) p. 7.
3. Riemer, 2014, p. 32.
4. Laura Gegget, "Ancient Knife-Toothed Reptile is Crocodile Cousin," LiveScience web site, January 21, 2015, https://www. livescience. com/ 49513-predatory-crocodile-fossil.html, paras. 1–3, 10.
5. Agata Blaszczak-Boxe, "Oldest Known Snake Fossils Identified," LiveScience web site, https://www.livescience.com/49582-oldest-snake-fossils-identified.html, paras. 1, 3, 8, 10.
6. Henry M. Morris, *Scientific Creationism* (Green Forest, AR: Master Books, 1985) pp. 78–90.
7. Donald Prothero, "Evolution: The Fossils Say Yes!" The Panda's Thumb web site, November 15, 2007, https://pandasthumb. org/ archives/ 2007/ 11/evolution-the-f.html, para. 3.
8. "List of Transitional Forms," Rational Wiki web site, September 16, 2017, http://rationalwiki.org/wiki/List_of_transitional_forms, para. 1.
9. Robin Lloyd, "Fossils Reveal Truth about Darwin's Theory," LiveScience

web site, February 11, 2009, https://www. livescience. com/3306-fossils-reveal-truth-darwin-theory.html, paras. 2, 5.
10. Janabi, Interview with Jerry Coyne.
11. Luther Sunderland, *Darwin's Enigma: Fossils and Other Problems* (Green Forest, AR: Master Books, 1988), p. 89.
12. Neil Shubin, *Your Inner Fish* (Episode 2) – "Your Inner Reptile," https://www.youtube.com/ watch?v=XxfnOBlEZX4, 0:04, 0:56.
13. Marvin Lubenow, Bones of Contention (Grand Rapids, MI: Baker Books, 1992), p. 72.
14. Ibid., p. 21.
15. "Evidence of Common Descent," Wikipedia web site, September 25, 2017,https://en.wikipedia.org/wiki/Evidence_of_common_descent, paras. 6–7. Evidence from comparative physiology and biochemistry/ Genetics quote #1 para. 1, quote #2 para. 2, Quote #3 para. 3., Evidence from paleontology/Fossil record quote #4 para. 4.
16. R. K. Sepetjian, "The Geologic Column: Invented to 'Free the Science from Moses,'"Across the Fruited Plain web site, October 11, 2011, https://sepetjian.wordpress.com/2011/10/11/the-geologic-column-invented-to-%E2%80%9Cfree-the-science-from-moses%E2%80%9D/, para. 8.
17. "Scientific Evidence," Wikipedia web site, September 15, 2017, https://en.wikipedia.org/wiki/Scientific_evidence, para. 2.
18. Hilary Hanson, "Humans Were Likely Boinking Neanderthals Earlier Than We Thought," *HuffPost Science* web site, February 17, 2016, http://www.huffingtonpost.com/entry/neanderthal-human sex_us_56bf63 a2e4b0b40245c-6dede, paras. 2–3.
19. Ruth Tennen, "Chatty Red-Haired Neanderthals?" The Tech Museum of Innovation web site, November 9, 2007, http://genetics.thetech.org/ original_news/news67, paras. 1–2.
20. "Neanderthal," Wikipedia web site, September 23, 2017, https://en. wikipedia.org/wiki/Neanderthal, paras.
21. "Cro-Magnon," Wikipedia web site, September 26, 2017, https://en. wikipedia.org/wiki/Cro-Magnon, paras.
22. "Brain Size," Wikipedia web site, September 15, 2017, https://en. wikipedia.org/wiki/Brain_size, para. 18.
23. "Neuroscience and Intelligence," Wikipedia web site, September 8, 2017, https://en.wikipedia.org/wiki/Neuroscience_and_intelligence#Brain_size, para. 5.
24. "Brain Size," WikiVisually web site, n.d., https://wikivisually.com/wiki/Cranial_capacity, para. 2.
25. Dennis O'Neil, "Neandertals," Palomar College web site, n.d., http:// anthro.palomar.edu/homo2/mod_homo_2.htm, para. 5.
26. Dr. Niels Harrit, "9/11 and the Seventh Tower" [interview with Niels Harrit, University of Copenhagen, Denmark, February 20, 2013], Age of Truth TV, http://ageoftruth.tv/dr-niels-harrit-911-the-seventh-tower/ 6:16, 36:32.

27. Bob Proctor, *You Were Born Rich* (Amarillo, TX: TAG Publishing, 2014), p. 109.
28. Alfred Rehwinkel, *The Flood in The Light of The Bible, Geology, and Archaeology* (St Louis, MO: Concordia, 1951), p. XVIII.
29. Rehwinkel, p. XVIII.
30. Henry M. Morris, a great scientist from our generation, earned a Ph.D. in hydraulic engineering (1950), held the position of professor at a number of universities. He co-authored an advanced textbook on engineering hydraulics that was used in dozens of universities worldwide. He also authored numerous books and helped found a number of Christian ministries. In his short book, *Men of Science, Men of God* he gives a short summary of 111 great scientists who made major scientific contributions and who believed in the Bible.
31. Jason Thibeault, "Abiogenesis is Not Spontaneous Generation, Period," The Orbit web site, June 25, 2010, https://the-orbit.net/lousycanuck/ 2010/06/25/abiogenesis-is-not-spontaneous-generation-period/, para. 3.
32. Kara Rogers, "Abiogenesis," *Encyclopedia Britannica* web site, n.d., https://www.britannica.com/science/abiogenesis, section 4.

> In fact, evolution became in a sense a scientific religion; almost all scientists have accepted it and many are prepared to 'bend' their observations to fit in with it.*

CHAPTER 6

LET'S GO SNIPE HUNTING

During my early teen years, I belonged to a Boy Scout troop for which my father was one of the Scoutmasters. We would go on camping trips to various places and participate in different events and activities. During one such camping trip, our Scoutmasters decided to have a snipe[1] hunt.

Our whole troop of teenagers was totally naive as to what a snipe hunt was. Our Scoutmasters described the snipe as small, flightless birds that came out only at night. They were seldom seen, somewhat rare, very cautious, and moved like lightning through the underbrush. They could be caught only if frightened or disoriented by loud noise. We were in luck, for the place we were camping was supposedly prime habitat for this mysterious, ground-dwelling, chicken-sized creature.

The scoutmasters never told us what we were supposed to do with any of the birds we managed to catch, but it was exciting; an adrenalin rush, for we got to run around wild in the fields and woods in the dark with just flashlights, sacks, and a few tin pans while we hunted the elusive quarry. Afterwards, we were told, there would be a campfire, games, singing, storytelling, soda and snacks for everyone. We were organized into two groups; one group would be crouched down in the tall grass and bushes and remain quiet and still (fat chance ... how do

---

*H. S. Lipson, FRS (Professor of Physics, University of Manchester, UK), 'A physicist looks at evolution'. *Physics Bulletin*, vol. 31, p.138.

you keep excited teenage boys quiet and still for half an hour?) with cloth sacks ready to carry the birds we caught running through the woods. The birds would be running to escape the commotion caused by the other group stationed a short distance away. Those boys would slowly walk toward us yelling and waving their flashlights all around while beating cooking kettles, grass, and bushes with sticks.

Our excitement was fueled by sporadic calls and shouts from the Scoutmaster's: "I think I see one!" "Over there!" "There it is!" "I think it's coming toward you!" I recall also hearing muffled laughter coming from where the Scoutmasters were observing the show we were putting on for them. We suspected that the hunt was a prank, but we were not sure, so we went along with it and had a great time. Some of us even thought we saw a fleeting shadow a time or two of this cryptic fowl.

The evolutionist's task has turned out to be a snipe hunt, a fruitless undertaking—a fool's errand. For many, it is a life-long endeavor, a full-time wild-goose chase. Numerous are those who have embarked upon its impossible search, living out the script of a real-life Mission Impossible. The only difference is that this mission will never come to an end; the snipe does not exist, nor will they find the beginnings of life as they envision them.

Whereas a wild-goose chase may be accidental, a snipe hunt is usually initiated as a prank. It is a type of practical joke played upon gullible, inexperienced, and naïve people. In the case of evolutionism, it is neither a prank nor practical joke, for those teachers and professors who espouse this doctrine purposely pass its misinformation to their understudies, protégés, and acolytes. What a horrific, contemptible, and senseless waste of time, talent, money, and resources searching for missing links in an imaginary family tree of ancestors. It is like researching the reason Rudolph can lead Santa's team at breakneck speed through the nighttime skies once a year.

INFAMOUS EVOLUTIONARY HOAXES

Why should we believe the findings of researchers, professors, and teachers who embrace, espouse, and champion evolutionism? Down through the years, in the name of scientific advancement, numerous frauds and hoaxes have been perpetrated upon the unsuspecting public by evolutionists. Many of these lies have been placed within textbooks that are used by schoolchildren. So as not to belabor the point, I will present just a few of the more infamous deceptions.

Piltdown Man is perhaps the most infamous hoax ever perpetrated. It was the "discovery" of a few bone fragments, parts of a skull and jawbone that were supposedly collected from a gravel pit at Piltdown, East Sussex, England in 1912. This evidence of evolution was used to promote the supposed evolution of man from lower forms of modern humans. In the 1925 Scopes Monkey Trial, the fossil was even introduced as evidence by defense attorney Clarence Darrow in his defense of science and math teacher John Thomas Scopes. It is very significant, for more than 40 years elapsed between its discovery and its full exposure in 1953 as an act of forgery; it turned out to be a modern human skull and an orangutan jaw. During most of that time, it was presented in children's textbooks as proof that humans had descended from lower forms of life. A vast expenditure of effort and time was spent on the fossil and more than 500 articles[2] and memoirs are said to have been written about Piltdown man.

Sometimes an accused person can have his or her whole life destroyed by just the accusation that they are guilty of some crime, so while it is true that scientists did eventually expose this hoax, it was way too late, for the damage had already been done. This evidence had already been paraded before a whole generation of schoolchildren, which convinced many that their great-grandparents crawled out of a slimy tide pool sometime in a distant, hoary, antediluvian epoch.

Nebraska Man is another memorable display of dishonesty among many paleontologists. In this case it was a *pig's* tooth that made a monkey out of evolutionists. In 1917, Harold Cook, a rancher and geologist from Nebraska, unearthed just one tooth, a molar from Pliocene deposits in western Nebraska. The tooth ended up in the hands of Dr. Henry Osborn of Columbia University, head of the American Museum of Natural History. In 1922, Osborn introduced *Hesperopithecus haroldcookii* to the world, the first anthropoid ape from America, a missing link in the line of human evolution. Newspapers such as the *Illustrated London News* published stories complete with pictures of Nebraska Man and his wife, all nicely reconstructed by an artist's brilliantly imaginative genius—all drawn from just the tooth of a pig.

Other learned men of science also sided with Osborn. Sir Grafton Elliot Smith, F.R.S., Professor of Anatomy of Manchester, England stated, "I think the balance of probability is in favour of the view that the tooth found in the Pliocene beds of Nebraska may possibly have belonged to a primitive member of the Human Family."[3]

Although not all were taken in by this misidentification, its quick acceptance by many shows how deeply saturated the minds of evolutionists had already become. During the summers of 1925 and 1926, further field work on the original site uncovered other parts of the skeleton and revealed that the tooth belonged neither to an ape nor a man but to an extinct species of peccary. In 1927, the journal *Science* retracted its earlier identification of the tooth as belonging to an ape-man.

The only so-called experts fooled by the pig's tooth were evolutionists. There is an antidote to rash and foolish evolutionary claims: faith in the Creator-God of the Bible.

Recapitulation theory was a devious fraud perpetrated by Ernst Haeckel (1834–1919), a German biologist, naturalist, philosopher, physician, professor, and artist. He was a fervent promoter of evolutionism and helped to popularize Charles Darwin's work in Germany. His theory claimed that an organism in its biological development parallels and summarizes its species' evolutionary development in the womb. He supported the theory with drawings of embryos which have been shown to be highly inaccurate due to his evolutionary beliefs. In reality, for someone with his artistic talent to totally misrepresent how developing embryos look amounts to fraud.

Haeckel's drawings vastly overemphasized the similarities between certain vertebrates and claimed this was evidence that fish and humans had evolved from a common ancestor. His original drawings were shown to be fraudulent well over a hundred years ago, yet those *same* drawings can still be found within books, magazines, and textbooks used by all grade levels to promote evolutionism.[4]

Another whopper used to promote evolutionism is the myth that during one stage in embryological development the baby has what is called *gill slits*, which are similar to gill structures in fish; which in turn supports the idea that humans share a common ancestor with fish. The folds or wrinkles in the baby's neck region are not gills, for they never have anything to do with breathing. Those slits develop into the inner ear, thymus, parathyroid, tonsils, and into a portion of the face. And yet, this myth is still found in textbooks used in high school and college biology as "scientific evidence" for evolutionism. Even in the popular book *Baby and Child Care* by Dr. Spock he claims that "as the baby lies in the amniotic fluid of the womb, he has gills like a fish." I was taught that myth when I was in school, were you?

> Ancestral characters are often, but not always, preserved in an organism's development. For example, both chick and human embryos go through a stage where they have slits and arches in their necks like the gill slits and gill arches of fish. These structures are not gills and do not develop into gills in chicks and humans, *but the fact that they are so similar to gill structures in fish at this point in development supports the idea that chicks and humans share a common ancestor with fish.* Thus, developmental characters, along with other lines of evidence, can be used for constructing phylogenies.[5] (emphasis added)

This fable can be found in the 1977 edition of the Encyclopaedia Britannica:

> Embryology was seen in an evolutionary light when the German zoologist Ernst Haeckel proposed that the epigenetic sequence of embryonic development (ontogeny) repeated its evolutionary history (phylogeny). Thus, the presence of gill clefts in the mammalian embryo and also in less highly evolved vertebrates can be understood as a remnant of a common ancestor.[6]

This myth can still be found throughout the Internet. Of course, one can also find "evidence" for alien abductions. However, one will never find a legitimate site for evolution, which offers or provides any real scientific evidence. On March 23, 2015, I found the website *Evolution: Frequently asked Questions*. In an article by Richard Peacock, titled "Five Proofs of Evolution," "proof" number four states:

> 4. Common traits in embryos. Humans, dogs, snakes, fish, monkeys, eels (and many more life forms) are all considered "chordates" because we belong to the phylum *Chordata*. One of the features of this phylum is that, as embryos, all these life forms have gill slits, tails, and specific anatomical structures involving the spine. For humans (and other non-fish) the gill slits reform into the bones of the ear and jaw at a later stage in development. But, initially, all chordate embryos strongly resemble each other.
>
> In fact, pig embryos are often dissected in biology classes because of how similar they look to human embryos. These common characteristics could only be possible if all members of the phylum *Chordata* descended from a common ancestor.[7]

At times, our long-lost tails are also referenced, along with mentions of our gill slits. Sometimes I wish had a tail. Imagine how useful it would be after grocery shopping; coming to your door with both arms full of grocery bags, you could use your tail to open the door.

How ignorant and regressive is the reasoning of evolutionists, for

Haeckel's fraudulent drawings are still being referenced as proof, as evidenced by his statement that human embryos look so similar to all the other members of the phylum Chordata.

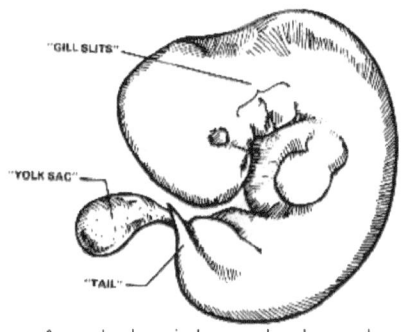

An early stage in human development
Figure 1. *What is Creation Science?* Master Books, 1982, p. 62

Are those common characteristics only possible because we have a common ancestor? Have the evolutionists ever considered the fact that anything we may have in common with other living creatures is due to our common designer?

Unbelievably, from those who should know better, Haeckel's recapitulation myth (while Haeckel might not be the one who originated this myth, he is well known for promoting it) is still being fed to students by the National Academy of Sciences (NAS) 1998 guidebook, *Teaching about Evolution and the Nature of Science*, made available to educators throughout America. Through the myth is a bit diluted in chapter one, "Why Teach Evolution" it states:

> The second question involves the inverse of life's diversity. How can the similarities among organisms be explained? Humans have always noticed the similarities among closely related species, but it gradually became apparent that even distantly related species share many anatomical and functional characteristics. The bones in a whale's front flippers are arranged in much the same way as the bones in our own arms. *As organisms grow from fertilized egg cells into embryos, they*

pass through many similar developmental stages. Furthermore, as paleontologists studied the fossil record, they discovered countless extinct species that are clearly related in various ways to organisms living today. (emphasis added)[8]

THE PEPPERED MOTH AND NATURAL SELECTION

Another mythological confirmation for evolutionism still found in most textbooks and Internet sites is the English moth, *Biston betularia*. This insect is frequently cited as an example of observed evolution. However, this story is one of the most pathetic and ridiculous attempts to hoodwink, deceive, and downright lie to students through their textbooks.

This ploy we will deal with in the confidence game was penned by Chris Colby, a writer for the *TalkOrigins Archive*. He writes the following:

> The English moth, *Biston betularia*, is a frequently cited example of observed evolution...In this moth there are two color morphs, light and dark...dark moths constituted less than 2% of the population prior to 1848. The frequency of the dark morph increased in the years following. By 1898, the 95% of the moths in Manchester and other highly industrialized areas were of the dark type. Their frequency was less in rural areas. The moth population changed from mostly light colored moths to mostly dark colored moths...
>
> The increase in relative abundance of the dark type was due to natural selection. The late eighteen-hundreds was the time of England's industrial revolution. Soot from factories darkened the birch trees the moths landed on. Against a sooty background, birds could see the lighter colored moths better and ate more of them. As a result, more dark moths survived until reproductive age and left offspring. The greater number of offspring left by dark moths is what caused their increase in frequency. This is an example of natural selection.[9]

Before we delve into an examination of this proof for evolutionism, we need to understand a little about *natural selection*, what it is and what it actually does.

Natural selection, along with mutation, migration, and genetic drift, are the supposed keys to the basic mechanisms of evolution. The only problem with these "keys" is they do not open any doors. Natural selection is a conservative, not a constructive process, for new information is never added or acquired. It is a process that selects from traits and variation already inherent and existent within a genus or species. Over time the most suitable features selected will become predominant in that life form and will be passed down generation to generation. However, those changes will never alter the genus beyond its own predetermined limits.

Natural selection is kindred to "quality control" on an automobile assembly line. That process weeds out damaged, marred, broken, impaired, and inferior parts and products. If the assembly line at the Ford plant was producing classic Mustangs, how long would it have to run until the Mustang vehicle was able to transform into a Cobra helicopter or a John Deere tractor? Nature in some respects is more permissive and less exacting than quality control at a factory; it will permit sight changes to occur—it will tolerate numerous varieties, colors, shapes, and sizes. However, all changes will be confined to specifications—to the programing and information that already exists within the DNA of the organism.

Over many generations, these slight changes may anatomically alter the organism, allowing it to adapt to a different ecological niche. If isolated long enough from others of its kind, it may become another variety of the same kind—what might be called a different species. Of course, the word "species" is an arbitrary and subjective term. For example, though brown bears and polar bears are thought of as different species, they can still interbreed and produce fertile offspring, for they are the same kind of animal.

These kinds of changes have been verified by intensive field research over the last quarter of the twentieth century, but no matter what kind of creature the studies started with, in the end, even after any kind of changes had been observed, the organism was still the same kind of creature. This is what Darwin described and observed during his five-week-long visit to the Eastern Pacific Ocean's Galapagos Islands. He identified 13 so-called species of finches, which he believed to have originated from the mainland of South America, nearly 600 miles to the east. These species of finches were slightly different from each other in beak size and shape, adaptions that allowed them to better obtain enough food to survive and reproduce. Thus, they were able to live in the different environments that were dispersed throughout the islands. However, this is not macroevolution, the transformation of one kind of living creature into a different organism, but microevolution—slight changes within an animal kind that produce varieties and diversity in organisms. And, lest we forget, they started out as finches, they were finches when Darwin studied them, and they are still finches 160 years later.

Now let us focus on the lesson that is still found in school textbooks: the *peppered moth*. Dennis O'Neil, in his article *Darwin and Natural Selection*, says this about the peppered moth:

> An example of evolution resulting from natural selection was discovered among "peppered" moths living near English industrial cities ... it is abundantly clear that there has been an evolution in peppered moth coloration due to the advantage of camouflage over the last two centuries ... [10]

This is the proof for evolution millions have been taught for decades, which I and my generation were taught in school during the sixties and which Colby and O'Neil write about. This key, as Darwin reasoned[11] is *natural selection*, the supposed basic mechanism of evolution, but what kind of evolution is it? Which of the two aspects of evolution, micro or macro, did we learn about with the example of the peppered moth? Neither! Natural selection did occur, but it changed nothing within the genes or DNA of the moth population; the *before* and *after* moths are identical.

At one point in time, it seemed as though more dark-colored moths were eaten by birds, but, for whatever reason, in time more light-colored varieties were consumed by the lurking winged predators. What happened within the moth population was just a swing in the number of colored moths, more one season and fewer the next. There was absolutely no change in the moths themselves, for both light and dark moths existed before the study and identical light and dark moths existed after the study. The question we need to ask ourselves is this: How long would it take for this process to change a single feature of the peppered moth, or change it into something else? If you said *never*, pat yourself on the back. You have seen the light.

Perhaps, in time, the light-colored moth population might be eliminated from the gene pool. Were that to happen, it would not be evidence of macro or microevolution, for the remaining moths would still be dark-colored moths, having incurred absolutely no change in physical coloring or DNA. What would have occurred is a loss of information within the gene pool; never again could the peppered moth produce a light-colored variety. Evolution, on the other hand, requires an increase of information in the gene pool. Thus, the evolutionary scenario is diametrically opposed to what is observed or needed!

In another work dealing with proof of evolution, there is a surprising admission by an evolutionist that states:

> "We do, I think have the right," [Stephen Jay Gould] wrote, "to be both astonished and ashamed by the century of mindless recycling [of the Haeckel drawings] that has led to the persistence of these drawings in a large number, if not a majority, of modern textbooks."[12]

If Gould was enlightened to the whole truth and was totally honest, his statement would have included every aspect, every nook and cranny of the numerous lies and fraudulent evolutionary beliefs found within textbooks.

You would never be able to reanimate a cup full of frog puree. You would never take a snipe to a taxidermist to be stuffed in order to put on display. You would never be able to take home a piece of Santa's sleigh for a souvenir, nor feed Rudolph a sugar cube. You will never see pigs fly with their newly evolved wings. All those things are nestled firmly within the realm of fairy tales, the same portion of a mind from where the lies and fiction of evolution springs forth.

If you still believe in evolutionism, are somewhat adventurous, and have a hankering to spend some time in the deep woods, I know a guide and expert in the field who works for a small fee; he specializes in hunting cryptic flightless critters that inhabit the forests and timberlands and come out only during cloudless summer nights when the moon is full.

NOTES ON CHAPTER 6

1. There are about 25 species of snipe—it is a small wading bird characterized by a very long slender bill—but what we were hunting does not exist.
2. "The Piltdown Man and 'Implements'," *Nature* 274(4419), July 10, 1954, pp. 61–62.
3. Christian Myers, *The Evolution Conspiracy: The Attack on Genesis* (Eight Winds Books, 2014).
4. Here are two examples, one from the text and one that uses an illustration and text.

> Scientists accept the theory of *biological evolution*—the change of one form of life into another. ... Some scientists think that certain steps in embryonic growth are similar to those that may have taken place during evolution. Thus, a mammal seems to go through successive embryonic stages in which it resembles a creature without a backbone, and then an animal with gills, and a backbone, before it finally become recognizable as a mammal. (*The New Book of Popular Science*, vol. 3, 2000 ed., Danbury, CT: Grolier, 2000, p. 418)

> Can you tell the difference between a chicken, a rabbit, and a human? It's pretty easy when you compare adults from each species. But what about comparing members of these species before they are born? Look at the left side of Figure 10, which depicts the very early embryos of a chicken, a rabbit, and a human.
>
> All the organisms shown in the figure are vertebrates, or animals

that have a backbone. Early in development, human embryos and the embryos of all other vertebrates are similar. These early similarities are evidence that all vertebrates share a common ancestor. Although the embryos look similar to each other in very early stages, none of them look like their adult forms. Embryo development has evolved over millions of years causing the embryonic structures to grow into many different species of vertebrates. (*Holt Science & Technology: Life Science*, Austin, TX: Holt, Rinehart, and Winston, 2001, p.183)

5. University of California Berkeley, "Learning about Evolutionary History," *Understanding Evolution* web site, n.d., http://evolution.berkeley.edu/ evolibrary/article/evodevo_02, para. 2.
6. *Encyclopaedia Britannica Micropaedia*, ed. William Benton (1943–1973, 1977) vol. 19, p. 1166.
7. Richard Peacock, "Five Proofs of Evolution," Evolution: Frequently Asked Questions web site, n.d., http://evolutionfaq.com/articles/five-proofs-evolution, para. 4.
8. National Academy of Sciences, *Teaching about Evolution and the Nature of Science* (Washington D.C.: National Academies Press), 1998), p. 1.
9. Chris Colby, "Introduction to Evolutionary Biology," The TalkOrigins Archive web site, January 7, 1996 http://www.talkorigins.org/faqs/faq-intro-to-biology.html, para. 5.
10. Dennis O'Neil, "Darwin and Natural Selection," Palomar College web site, n.d., http://anthro.palomar.edu/evolve/evolve_2.htm, para. 15.
11. Mayr, 2009.
12. Jonathan Wells, *Icons of Evolution: Science or Myth?* (Washington D.C.: Renergy Publishing, 2000), pp.108–9.

There was little doubt that the star intellectual turn of last week's British Association for the Advancement of Science meeting in Salford was Dr. John Durant, a youthful lecturer from University College Swansea. Giving the Darwin lecture to one of the biggest audiences of the week, Durant put forward an audacious theory—that Darwin's evolutionary explanation of the origins of man has been transformed into a modern myth, to the detriment of science and social progress, . . .

Durant concludes that the secular myths of evolution have had "a damaging effect on scientific research", leading to "distortion, to needless controversy, and to the gross misuse of science".[*]

CHAPTER 7
JUST ADD WATER AND SOME TIME

How am I doing so far? Are you beginning to understand the study of evolution is an exercise in futility, for it serves no useful propose? Are you aware that over the course of many decades, the study of evolutionism's doctrines has not produced a single useful irrefutable fact, historical information, or any pertinent scientific finding or advancement that would be of benefit to mankind? Of course, at times a bit of useful information may have come out of its study, but that would be in spite of, not because of.

---

[*]Dr John Durant (University College Swansea, Wales) as quoted in 'How evolution became a scientific myth', *New Scientist*, 11 September 1980, p. 765.

A PETITE PRIMATE NAMED LUCY

Muse over the time spent on research and promotion for Lucy, the fossil specimen discovered in 1974 in Ethiopia by paleoanthropologist Donald Johanson, curator of the Cleveland Museum of Natural History. About 40 percent of her skeleton was recovered. After the public announcement of this discovery, at the time, Lucy garnered so much public interest she became famous worldwide, and became a household name. Johanson published a book about the discovery and reconstruction. Starting in 2007, Lucy and a selection of associated artifacts made an extended six-year tour of the United States. The exhibition was titled *Lucy's Legacy: The Hidden Treasures of Ethiopia*.

The original skeleton has been preserved and numerous plaster replica skeletons have been made. The work of excavation of those fossils was very difficult. The preservation and analysis of those specimens was time-consuming work. The University of Texas at Austin performed high resolution CT scans of the fossils. Setting up the dioramas of these finds in various museums around the country took many hours of work. These displays present Lucy (*Australopithecus afarensis*) and other "human" predecessors showing each species, its habitats, capabilities, and behaviors. In the Field Museum in Chicago, as part of the diorama, a cast skeleton and a corpus reconstruction of Lucy (a supposedly 3 ft., 7" tall, 64 lb., chimpanzee-like beast) was displayed.

The initial attempts made in 1974 to date the fossils were hindered by several factors, one being the fact the rocks in the recovery area were chemically reworked or altered by volcanic activity. So, years later, in 1990 a more precise argon-argon technology was used by Aronson and Robert Walter in the geochronology laboratory of the Institute of Human Origins (that certainly sounds like a non-biased research faculty). Normally paleoanthropologists estimate the age of hominin fossils by where they were found in relation to the layers of volcanic ash, which in this case the older was found 18 meters below and the younger layer was only 1 meter below the specimen, and they came up with dates of 3.22 and 3.18 million years.

In Chapter 10 we will learn why it is impossible to date anything from the past using radioactive elements. For now, as we continue to press forward in our study, we will discover any information or findings concerning Lucy or other fossils is always biased, prejudiced, and slanted toward predetermined conclusions, all to one purpose and goal—to produce faith in the religion of evolutionism.

Consider the immense amount of time, money, talent, resources, and research spent on just this one example. Now multiply that time by countless other evolutionary "research" projects being conducted across the world. For example, the online Encyclopedia.com site's article, *Human Evolution*, gives us a glimpse into the multi-tentacle arenas of evolutionary studies:

> Human evolution is the lengthy process of change by which people originated from apelike ancestors starting nearly five million years ago. The modern scientific study of human evolution is called paleoanthropology. A subfield of anthropology, this discipline searches for the roots of human physical traits, culture, and behavior. It attempts to answer questions: *What makes us human? When and why did we begin to walk upright? How did our brains, language, art, music, and religion develop?* By approaching these questions from a variety of directions, using information learned from other disciplines such as molecular biology, paleontology, archaeology, sociology, and biology, *we continue to increase knowledge of our evolutionary origins.* (emphasis added)

Because evolution did not happen, Lucy is a chimpanzee-type creature, not a part of man's family tree. All evolutionary research is totally useless, for there are no evolutionary origins to study. All the questions evolutionists ask deal with imaginary events, thus, any research undertaken will never provide any factual historical information or knowledge which would have practical value or purpose.

The consequence of this state of affairs in the scientific and academic communities is extremely disturbing. Due to the study of evolutionism and its application, there is a massive drain of human capital and resources that would otherwise be channeled and directed to activities that would be of benefit to every nation. Is this not evil and reprehensible, when the world is in desperate need of solutions, real answers to numerous problems vexing mankind, it has been insidiously *robbed* of the intellect, perspicacity, and abilities of so many of our gifted young people, and redirected their resourcefulness, ingenuity, and imagination, to toil all their lives in fields of useless, valueless, and pointless endeavors?

For those reading this book who may be a proselyte of evolutionary philosophy, if you still have a somewhat open mind and a bit more spare time, during the next few pages we will expose a few more rotten spots and deficiencies of evolutionism. In so doing, I hope to direct you to a more productive and useful path of study and research—one that will not waste your talents, abilities, or precious resources, but will actually add to man's store of knowledge and be of benefit to humanity.

A ZIPPERLESS ZIPPER

Most people have heard of Velcro: it is a hook-and-loop fastener made of two strips of nylon and polyester fabric, one with thousands of tiny loops and another with tiny hooks. They "mate" when pressed together, attaching firmly but temporarily, and can then be pulled apart. There are already hundreds of known uses for this product and more are being discovered each year. It is used in a wide array of industries, including apparel, construction, agriculture, transportation, and packaged goods. It keeps rugs from slipping and floor mats in cars in place. It is used as fasteners for slipcovers, drapes, clothing, and on airplane seat cushions that are used as flotation devices. It is used on the band when you have your blood pressure checked and on hospital patient gowns. It is even said that Apollo astronauts used it to secure pens, food packets, and other equipment they did not want floating away.

In 1941, Swiss electrical engineer Georges de Mestral, also an inventor and amateur-mountaineer, went for a nature hike with his dog in the nearby Swiss Alps. They returned from their jaunt covered with burrs, plant seed-sacs that cling to fur and clothing. Curious as to why those seeds clung so viciously to the dog's fur and the fabric of his trousers, he examined the burrs under a microscope and discovered small hooks, which he determined had enabled those seeds to cling so tightly. As he was an inventor, he wondered if those burrs could be turned into something useful. After eight years of research, he successfully reproduced a unique fastener that mimicked what nature had already designed. He named his invention Velcro, a combination of the French words *velours* (velvet) and *crochet* (hook). He formally patented it in 1955. After subsequent refining and developing, its practical commercial manufacture as a *zipperless zipper* began in the late 1950s.

Many believe Velcro is the brainchild of Mestral, but he simply *mimicked* what nature had already designed. *Nature*, as I use the term, does not mean a mindless nothing, but refers to the cosmos and all it contains, which was created by the Lord Jesus Christ. He is the master designer, the original inventor and engineer of all we see around and within us. "*All things were made by him, and without him was not anything made that was made*" (John 1:3). Velcro is but an inferior copy of one of God's marvellous inventions.

It has been said that "we are always safe when we copy nature." Many of the products and technologies we see all about us are simply reinventions, copies, or mimicry of what God, in His wisdom, designed into the very structure, composition, and fabric of life in the beginning. Mestral did not invent Velcro; he just copied a design that was already in existence. However, he had to spend many hours in research, experimentation, and testing to reverse-engineer a structure which God had already perfected.

The bird in flight and the fish in the sea are well designed and perfectly adapted for the medium they inhabit. Often, scientists will look at nature, God's way of engineering, when they want to design a faster submarine, a more fuel-efficient airplane, a stronger substance, etcetera. For example, on February 19, 2015, Alyssa Newcomb posted an article titled *Strongest Stuff on Earth*. Sea snails' teeth, it seems, rival the strength of steel and bulletproof vests. Since this discovery, scientists have been working to create similar materials that can be used on everything from boats and aircraft to dental fillings. Asa Barber, a researcher who worked on one such study published in the *Journal of the Royal Society*, stated:

> These teeth are made up of very small fibers, put together in a particular way... *We should be thinking about making our own structures following the same design principles.*[1] (emphasis added)

In a March 2016 David Gutierrez, staff writer for *NaturalNews*, posted this story: "Mother Nature's Nanotechnology: Oak Leaf Turned into Functioning Battery." In the story he explained the ingenious method used to change an ordinary oak leaf into a key component of a high-storage, long-lasting rechargeable sodium battery:

> We have tried other natural materials, such as wood fiber, to make a battery," researcher Liangbing Hu said. "A leaf is designed by nature to store energy for later use, and using leaves in this way could make large-scale storage environmentally friendly.[2]

A leaf is designed by "Mother Nature"? Who is this great nothing, that mindless entity which only deals in blind chance? One which can solve complex engineering problems without scientists, programmers, researchers, or engineers? When scientists, farmers, and others do as suggested by Job, "to ask the beasts, the fowls, fish and the earth, for it shall teach thee" (Job 12:7–8), great discoveries are often made, but most of the time, the original designer, the creator of the thing being copied, is not given credit, though all the credit is due Him.

The prodigious scientist and teacher, Dr. George Washington Carver, spoke to the Creator often, and one time asked Him, "Please Mr. Creator, will you tell me why the peanut was made?" In time, God provided a splendid answer to his inquiry.

Dr. Carver asked that question of the Creator because he had encouraged sharecroppers to grow peanuts, for that crop would help replenish soil that was worn out from planting their fields year after year with cotton. However, to his dismay, after the harvest there was little market for those legumes. Therefore he went to work to enhance the market value for peanuts. He is credited with inventing over 300 uses for the peanut, using them in everything from soap to soup. Interestingly, it seems the Aztec culture made a paste from peanuts and Edson obtained a patent for "peanut paste" more than a decade before Carver began his work at Tuskegee Institute. However, the invention of peanut butter has been credited to at least three other individuals, Marcellus Gilmore Edson of Canada, Dr. John Harvey Kellogg (creator of Kellogg's cereal), and Dr. Ambrose Straub of St. Louis, Missouri.

THE INVISIBLE CARETAKER

God does not need man to water His forests or grasslands. Man is not tasked with fertilizing the woodlands. In places where man does not cultivate the soil, the Creator does not ask man to improve the ground by plowing or digging it up so organic plant material can be mixed into it. God has even designed fire to be a useful tool on his land. It helps keep His forests rejuvenated and fruitful by getting rid of old and diseased plants, clearing the way for younger and healthier trees and bushes to replace them. Fires caused by lightning are a common occurrence, and when it does not happen often enough, the underbrush and growth builds up to dangerous levels, so when a fire does occur it will burn not just the underbrush but will spread to the tree canopy, engulfing the whole forest and devastating it.

On April 19, 2015 sparks from an unattended cooking fire started a wildfire in California. This fire threatened hundreds of homes near the cities of Norco and Corona, about 35 miles southeast of Los Angeles. Captain Mike Mohler of the California Department of Forestry and Fire Protection said the fire burned in the Prado Dam Flood Control Basin, where the vegetation has not burned in decades. As a result, there was up to 3 ft of "duff"—needles, leaves and other combustible plant debris. Thankfully, no property damage or injuries were reported, for fire-fighters were able to contain it before got out of control, and I hope they have learned a lesson about using God's provision, fire, in a helpful way and to never allow this much combustible material to build to such dangerous levels again.

There are a number of plants and trees which need fire in order to release their seeds and survive. The jack pine, which is native to the extreme north-eastern and central parts of the North American continent, has cones that are very thick and hard. They are literally sealed with a strong resin, which only heat from a forest fire can melt to allow the cone to open and release the seeds. Similarly, the cones of the giant redwood[3] and sequoia trees of California, which can tower well over 300 ft and grow to 50 ft in diameter, can contain up to 200 seeds and take about two years to mature. After maturity they will remain sealed within the cone until the heat from a forest fire causes the cones to open and liberate their seeds.

The Native Americans and, to a lesser degree, the early settlers, had a significant effect on the North American continent through their use of fire. These nations of people understood the great benefit of fire, God's provision, and used it to modify the vegetation and create and maintain the diversity of ecosystems. Repeatedly using controlled surface burns every few years allowed them to keep large areas of forest and mountains free of undergrowth and small trees, thereby preventing uncontrolled fires that destroy everything in their path.

It seems the U.S. National Forestry Service has finally learned this. On May 5, 2015, I found this story: "Feds Expand Efforts to Fight Wildfires by Reshaping Forests." The article says:

> Other areas of the Coconino National Forest are *blackened intentionally by fire*, giving native plants a boost and any wildfire less of a chance to explode into something catastrophic.
>
> The complex wildfire prevention effort encompassing four national forests…aims to restore the forests to conditions ideal for wildlife, streams and cultural resources while creating a buffer for communities where large wildfires might devastate the landscape. […]
>
> *Returning parts of the forests to a time where trees were more spread out and fire naturally swept the landscape* has spread out and has been credited with helping save the towns of Alpine and Nutrioso in the 2011 Wallow Fire…and with keeping last year's Slide Fire from spreading into Flagstaff. […]
>
> Dick Fleishman, assistant team leader for the forest restoration project, pointed out grasses and brush that are rejuvenated by prescribed burns and the absence of low-lying limbs on trees that would allow flames to climb up top where fires burn hotter and more intensely. Many of the trees in the 600,000 acres approved for restoration make for an unhealthy forest because they are around the same age and will need to be cleared out, he said. (emphasis added)[4]

Besides God's provision for fire, His remarkable designs work well in every other function necessary for life. As example, look at God's method of water collection, storage and purification. In most places where man has not disturbed the rock layers by mining, drilling, hydrofraction (fracking), or other methods, rainwater seeps into the ground, through which it is filtered and purified before it enters large underground aquifers. These natural storage faculties replenish themselves and their waters stay pure. To access their waters, pumping is rarely even needed, for springs of water just percolate and flow up through the ground. The water comes out cool and uncontaminated by pathogens or other pollution. Whereas man in his wisdom needs to spend billions of dollars to build water purification plants.

Some contend that groundwater is insufficient to support the needs of the current human population's farming and plumbing needs—that even with supplementary purification of recycled water drawn from other sources, underground reservoirs are being badly depleted. I would, for the most part, concur with that assessment. However, in our next section, we will be exploring a method of agriculture, which I believe, if it was widely practiced, could be the solution to this problem.

THE BACK TO EDEN METHOD OF ORGANIC GARDENING

Paul Gautschi, organic gardener, arborist, and resident of the state of Washington, is learning about God's provision and has spoken to the God of the earth. He has asked Him questions about the growing of food, and has reflected upon these words written in the Scriptures:

> But ask now the beasts, and they shall teach thee; and the fowls of the air, and they shall tell thee: Or speak to the earth, and it shall teach thee: and the fishes of the sea shall declare unto thee. Who knoweth not in all these that the hand of the LORD hath wrought this? In whose hand is the soul of every living thing, and the breath of all mankind. (Job 12:7–10)

Gautschi has observed the way God looks after His earth, the methods He uses, how He grows and cares for the trees, forests, and grasslands. He has simply done his best to emulate the earth's ecologies, and has found great success. However, unlike Gautschi, evolutionists would like the rest of us to believe that through trial and error a mindless nothing created life, then organized and planned the water, ecosystems, ecologies, and natural balances, which work so efficiently that no outside help is needed. Of course, at times floods, earthquakes and other disasters take place, which if given enough time, would heal all by itself; but at times man should step in and lend a hand, for he is God's steward.

Gautschi's lifelong journey back to the simple, sustainable, productive methods of gardening—of growing wholesome organic foods—is documented in the film *Back to Eden.* Since the film's release in August 2011, it has been viewed online over 2.5 million times in 218 countries. It has been endorsed by numerous groups, agencies, and organizations. As the film opens, Gautschi relays this information:

> When you look at the incredible landscape on planet earth, all the different terrains, the varying soil conditions, the awesome water features, oceans, lakes, rivers, streams, the waterfalls, different climates, the huge amounts of plants and ground covers, the requirements are so varied, can one fathom how big a project that was? When God designed the landscape project for planet earth, He was so genius; He designed it in such a way He would never have to show up for work. It is completely self-sustaining.[5]

If you grow plants or have a vegetable garden, I urge you to get online and watch the film. There are also numerous other short videos in which Gautschi answers many questions and queries about his methods. I do not have the space in this book to lay out all the methods Gautschi uses, but I hope to relay enough information to arouse your curiosity so that you will want to learn more about this fantastic way of gardening or farming.

One of the key aspects of Gautschi's success is *the Covering*. Wherever God grows His plants and the ground has not been disturbed by a natural disaster such as a flood, or man, the ground will always be covered by leaves, twigs, stones, grasses, or forest litter. Those things act as a means of protecting the soil, to keep it in place and prevent it from being blown or washed away. Gautschi does not plough, till, or dig up his soil. Nor does he mix organic plant material into his ground. One of Paul's rules for starting your garden is *do not dig up the soil* where you intend to plant your garden. In fact, once you have everything prepared, you never dig, mix, or till your garden soil, EVER! Just lay or place more organic material on the top as needed.

The idea that you should never dig or till your soil took a while for me to accept. That is certainly not the way I have always gardened and is a new idea to me, and undoubtedly it will also be to most people. But that is how God gardens and Gautschi simply emulates His methods, through which he has had overwhelming abundant success. Even though the place where Gautschi lives only receives 14" of rain a year, he has not watered his vegetable garden in over 15 years. And though his apple orchard was planted 31 years ago, he has never watered it once. During some summers there have been long periods of time without rain, yet the plants do not wilt and the fruits and vegetables stay luscious and full of moisture. He does water newly planted seeds, but once they sprout, he never waters them again. He does not fertilize, nor does he need to use any kind of pesticides, fungicides, or weed killers. His soil never compacts, making weed removal very easy. His plants are seldom bothered by harmful insects, and neither slugs nor snails use his garden as their neighborhood salad-bar, a place for picnic gatherings or town-hall meetings.

Crops that supposedly need acidic or alkaline soils respectively, are flourishing, growing right next to each other in the same soil. Plants that need full sun are doing just fine in shade. Crop rotation is not necessary in his garden. He does not need to mound his potatoes, and crops that supposedly need substantial amounts of water, are growing in the same soil and same conditions as all the other plants and are thriving.

Gautschi has none of the problems normally associated with "modern" man-made farming mono-culture techniques and agricultural methods. However, there may be one problem that those who use his methods might encounter when starting a garden in depleted and worn-out soils—a lack of trace minerals. There are a number of ways this can be remedied, such as using rock dust. Once the soils' trace minerals have been restored by minerals from the proper sources, this initial application will not have to be repeated, for those minerals and nutrients will be constantly replenished by the decomposing woodchips, vegetable scraps, and other organic matter you periodically reapply to the top of the ground.

Using Gautschi's methods, which are patterned after the way God cares for His land, it makes no difference what kind of soil you start with—sandy, hard-pack, clay, rocky, or even no soil at all. Even if the only space you had to plant a garden was a rock-hard gravel parking lot or driveway, within a few years you would be harvesting nutritious produce from those very spots without removing any small rocks or digging up the ground first! Because his methods work, growing great vegetables, herbs, plants, bushes, or trees is possible anywhere—be it rainforest or desert! However, realizing the full effect of using his methods will normally take time.

How does someone start this kind of garden? First, lay down a layer of cardboard or newspaper on untilled ground. This helps keep any weeds or grass from growing up through the soil, and it decomposes after a while. Next, lay a generous 2" layer of compost directly on the newspaper and a 4" to 5" layer of wood chips on top of that. That is it! If the compost is fully aged, you should be able to plant your garden right away (plant the seeds into the composted soil, not the woodchips). Remember, never dig up your soil, just spread more organic material on the top.

The kind of woodchips he uses are from chipped branches of bushes and trees, which includes all the needles, flowers, leaves, and twigs, not just bark, for you need the green part of the plant to produce good soil. One good source of organic material (fertilizer) comes from what Gautschi calls his soil manufacturing plant, the small number of chickens Paul keeps in a coop. He never feeds them grains, just the peelings and cuttings from his garden. After the chickens have dug through it for a while and it has been thoroughly aged, he sifts it and spreads the newly created soil—the droppings from his chickens—all over his garden as another source of nutrients.

You may be wondering, using Gautschi's methods of farming, how do the necessary levels of primary plant nutrients such potassium, phosphorous, and nitrates in his soil compare with the recommended levels? Reflect on these results:

> A recent soil test in Paul's garden revealed these results: "Listen to these numbers," Paul says. "On the test, you get two lines—the desired level that you want, and your lab results. The nitrates: the desired level was 40; my lab result was 120. Phosphorous, the desired level is 174; mine is 2,345. Potassium, the desired level is 167; mine is 1,154. Coming down to the smaller numbers: zinc, the desired level is 1.6; mine 21.5. What I love about this is I didn't do anything!"[6]

When I started gardening using Gautschi's methods I did not have access to woodchips or composted material, so I used whatever organic material was at hand. I use what some call the *chop and drop* method. What I prune from my plants and trees, I just chop up and drop at the foot of the plants. Around all my plants and flowers I just lay down the weeds I pull out and chop up, small chopped-up branches, grass, leaves, and lots of fresh banana peels. For my compost pile, I place larger branches that I have chopped up by hand, tea bags, leaves, vegetable, and fruit peelings. Some say not to compost citrus fruits, weeds, bread products, or rice in the pile; I say, to make things simple, other than no meat, human, or animal waste, if it rots, in it goes. If the weeds you threw in the pile start to grow pull them out, chop them up and throw them back in. [The reason some instruct gardeners not to compost meat, fish, and animal waste, it attracts vermin. Things like macaroni and cheese, oils, and bread take a long time to decompose and can create odors. Composting is just a way of getting organic material to rot faster. Nature composts everything ... why not do the same?] I do not worry about the pile's temperature, covering it, turning it over, or getting too wet or dry. Because it is always warm where I live, and there is plenty of rain, no matter what I do, in about three or four months, on the bottom of the pile there will be freshly composted, rich, black top soil.

Around the plants, in places where the mulch is now about 4" deep or so (I am always adding more) and has been decomposing for about four months, the soil beneath the mulch, which was originally just sand and gravel, is literally being transformed, almost as by magic, into rich dark soil, complete with all kinds of life, worms, insects and different kinds of critters that all make themselves right at home; all this without once digging or tilling the ground! The plants are becoming a rich green color, thriving, a good example of what healthy vegetation should look like. Gautschi's gardening techniques are not unique to him, nor is he the first one to use these kinds of methods. Similar methods have been used successfully around the globe in the past, and are still being observed today. However, in large part due to the film, numerous folks are now starting to take notice and apply his systems, with great success to their gardening efforts.

Ernst Gotsch is a Swiss researcher and farmer who migrated to Brazil in the early 1980s and settled on a farm in the cocoa zone of southern Bahia. His approach is called *agroforestry* and its goal is sustainable *farming in the forest*, which is an attempt to mimic the natural regeneration of forests. Starting with completely degraded soils, within a short time (five to eight years) he was producing above-average yields of cacao. Through soil recovery techniques very similar to Gautschi's he has remarkably increased the biodiversity of 480 hectares of land. Like Gautschi, Ernst Gotsch is using God's technologies and His environmental systems to increase productivity and generate sustainable agriculture in Brazil. His *sintropic* agriculture uses no irrigation, produces its own fertilizers, and the soil continues to improve. After observing the effects his methods have had on the land, Gotsch stated: "It [assures] me I am on the right path: *working to create areas of permanent inclusion of humans instead of areas of permanent protection from humans.*"[7] Gotsch's statement is a clear repudiation and rejection of man's ideas, with their environmentally destructive farming methods and evolutionary ideology. Modeling God's methods always brings good results.

A WHOLE LOT OF SHAKING GOING ON

The earth as designed by God is self-sustaining, self-healing, and normally does not need outside intervention from man. However, at times, accidents, mishaps, and catastrophes do occur and man's help is sometimes necessary.

Located in the southern part of the State of Washington is the beautiful Cascade Mountain Range. It is here where Mount St. Helens is located, nicknamed the "Fuji-san of America," because of the extensive snow and ice cover and exquisite symmetry of its pre-1980 summit cone.

After months of numerous small quakes, on May 18, 1980 an earthquake measuring 5.1 on the Richter scale struck below the north face of this majestic volcano and triggered the largest landslide in recorded history. The slide removed 1,300 ft off the top of the volcano. This allowed pent-up pressure to be released, and caused a lateral blast with pyroclastic flows, glowing clouds of superheated gas and rock debris. These flows moved at nearly supersonic speeds, hugging the ground as they cascaded across the surrounding landscape, flattening huge majestic century-old Douglas fir trees like rotten branches while the 1,800 degree heat of the flow incinerated almost everything in its path. Everything within eight miles of the blast was wiped out within a few minutes, and for almost 400 square miles there was massive damage. The intense heat from the eruption rapidly melted the mountain's snow and glacier covering, which generated massive lahars. The blast was heard for hundreds of miles. Fifty-seven people lost their lives, and over 4 billion board feet of usable timber, enough to build 150,000 homes, lay prone all across the landscape, all their trunks neatly aligned to the north.

In keeping with the evolutionary belief that man is not needed, in 1982, environmentalists were able to convince Congress to create an 110,000-acre Mount St. Helens National Volcanic Monument.[8] This monument is a parcel of ground that has been set aside, a hands-off domain where man was not allowed to help the healing process, so the land was left to respond naturally to the disturbances. There, the useable lumber had been left to rot; hillside erosion was left unchecked, and after 36 years the forests are still in the early stages of their struggle to recover.

In the rest of the blast area lands outside of the monument, for the most part, private, state, and national forest recovery was swift and dramatic, for within five years after the eruption an extensive logging salvage program was carried out and a great deal of replanting was accomplished. You will now find plantations of huge Douglas fir and other trees growing thickly in many places. Man's restoration and conservation efforts really paid off! It seems that most so-called environmentalists have an evolutionary worldview that man is a latecomer to this planet. He is not needed, and should, for the most part, take a hands-off approach to nature and the natural environment. This is clearly wrong, as stated before, man is to superintend this Earth, and when needed, to extend a helping hand. Many tree-huggers consider mankind a pest, cancer, or virus that needs to be eliminated, or at least greatly reduced in population.[9]

A January 22, 2013 story by Louise Gray was titled "David Attenborough—Humans Are Plague on Earth." Whether one agrees with or wholeheartedly loathes Attenborough for some of his political views or other reasons, many still consider him a patron of population matters. According to him:

> The television presenter said that humans are threatening their own existence and that of other species by using up the world's resources.
>
> He said the only way to save the planet from famine and species extinction is to limit human population growth.
>
> "We are a plague on the Earth. [...] Either we limit our population growth or nature will do it for us..." [...]

> Sir David...has spoken out before about...the need for investment in sex education and other voluntary means of limiting population in developing countries.[10]

Numerous are the voices of the ideologues that mimic Attenborough's evolutionary sentiments. James Lovelock spearheaded the scientific study of global warming. He is a scientist, environmentalist, futurist and author of ten books. His recent 2009 volume is entitled *The Vanishing Face of Gaia: A Final Warning*. In it he claims that humanity is "Earth's infection." Professor emeritus of journalism at the University of California, Berkeley, A. Kent MacDougall had a 25-year newspaper reporting career. He wrote a piece on population problems titled *Humans as Cancer* that can be found on the Internet.[11] At the end of his short story is a long list of references, most in the same vein. The moguls of Hollywood echo the same sentiments. In 1999 the Wachowski siblings created the enormously popular science fiction action film *The Matrix*, starring Keanu Reeves, Laurence Fishburne, and Carrie-Anne Moss. Their vision of a dystopian world brought to life was impressive. Though the emphasis of the story appears to be to save humanity, the real theme is that man is the savior: Neo is the "one" that will save what is left of mankind. The Wachowskis' later movie *V for Vendetta* revisits the same theme: one man alone will make things right, no God is needed.

In *The Matrix*, the human race had been enslaved by intelligent machines to be used as an energy source. Their world was actually a simulated dream-world reality called "the Matrix". Neo, a computer programmer, learns this truth and is drawn into a rebellion against the machines with other people who had been freed from the dream-world. At one point, a sentient computer program known as Agent Smith confronts the hero Neo, and relays a key point in the movie, that humans are "functionally equivalent in ecological terms to a virus."[12] Agent Smith's monologue goes as follows:

I'd like to share a revelation that I've had during my time here. It came to me when I tried to classify your species and I realized that you aren't actually mammals. Every mammal on this planet instinctively develops a natural equilibrium with its surrounding environment, but you humans do not. You move to an area and you multiply and multiply until every natural resource is consumed, and the only way you can survive is to spread to another area. There is another organism on this planet that follows the same pattern. Do you know what it is? A virus. Human beings are a disease, a cancer of this planet. You are a plague. And we are the cure.[13]

On the Internet site *The Teeming Brain*, Matt Cardin posted an article titled *It's Official: The Human Race is Earth's Disease*. He wrote:

Variations on the idea of humans as a disease on the planet and/or of human consciousness as an alien and destructive development have appeared in science fiction and horror fiction for decades. And the idea that the human race may be a destructive species without whom planet earth would be better off, and regarding whom planet earth may be prepared to take decisive cleansing action, has wound its way through the radical environmentalism movement since its birth in the 1970s...the widely quoted assertion by Paul Watson—militant whale protector, founder of the Sea Shepherd Conservation Society, and an early member of Greenpeace—that "Humans are presently acting upon [the earth's ecosystem] in the same manner as an invasive virus with the result that we are eroding the ecological immune system. A virus kills its host and that is exactly what we are doing with our planet's life support system. We are killing our host the planet Earth.[14]

Before we leave this topic, to give our discussion a little more spice, let's consider a somewhat bizarre and incongruous assembly of folks who call themselves The Church of Euthanasia. This organization was founded in Boston, Massachusetts in 1992 by "Reverend" Chris Korda and Pastor Robert Kimberk. According to Korda, it is likely that this group is the world's only anti-human religion. Their website states that they are "a non-profit educational foundation devoted to restoring balance between Humans and the remaining species on Earth." This statement gives the impression that this is a benign and innocuous bunch of folks who want to help humanity. However, for this "anti-human" group, there are four pillars of faith: suicide, abortion, cannibalism, and sodomy. They have just one commandment: "Thou shalt not procreate." Its most popular slogan is "Save the Planet, Kill Yourself." Sounds like they are asking for volunteers. May I suggest and hope that they are the only ones who will volunteer for their ignoble cause.

Just in case you were wondering if there are others of this ilk, these comrades do not stand alone; they are linked to several other "anti-humanist" organizations such as the Voluntary Human Extinction Movement and the Gaia Liberation Front, whose philosophy is summed up as "the Humans must be completely exterminated, ASAP." These cults are the logical end result of our death-obsessed, evolutionary influenced culture.

I can understand some of the reasons for Attenborough's assessments and conclusions. As I read through MacDougall's narrative, *Humans as Cancer*, he stated facts, made some valid points and used arguments that were in some cases seemingly irrefutable. However, because the philosophy and dogmas behind their analyses and ratiocinations is faulty, they all come to the wrong answer, solution, and resolution. While many evolutionists would probably find fault with the Church of Euthanasia, both groups are standing upon the same foundation of error, evolutionary beliefs, which will never lead to a satisfactory solution. Contrary to evolutionary consensus and assessment, humans are an integral, vital, and fundamental part of this planet. Not a pest, nuisance, or evil entity, even though at times this seems to be the case.

This planet was created by God as a place for man. *"...and let them have dominion over the fish of the sea, and over the fowl of the air, and over the cattle, and over all the earth, and over every creeping thing that creepeth upon the earth"* (Genesis 1:26). *"What is man, that thou art mindful of him?...For thou hast made him a little lower than the angels, and hast crowned him with glory and honour. Thou madest him to have dominion over the works of thy hands; thou hast put all things under his feet"* (Psalm 8:4–6). See also Genesis 9:1–7 and Hebrews 2:6–8. Even before sin entered into the world, labor was an integral, fundamental, and basic element of life for man, because employment at some craft or business contributed to man's happiness, the make-up of his body as well as his mind. This seems to prove that man was never intended to live just a contemplative life, filled only with pleasure, for only meaningful work will truly bring fulfilment to a life.

God placed mankind on this planet in a garden and gave us tasks, responsibilities, and activities to consume a portion of our time. *"And the LORD God took the man, and put him into the Garden of Eden to dress it and to keep it"* (Genesis 2:15). As shown by Mount St. Helens and Paul Gautschi's and Ernst Gotsch's examples, when man fulfills his responsibly of stewardship as God intended over this planet, this earth is better for it, than without him.

After the film *Back to Eden* came out, many farmers came to visit Paul to ask him questions: "You don't use pesticides or poisons, so why don't you have slugs, aphids or other harmful insects eating your plants like we do?" "Why don't you have any mud in your garden?" "You only get 14" of rain a year where you live, so why don't you ever have to water your garden?" Others ask: "How do we save the planet?" The answers to these questions are simple, whether you are inventing Velcro, a new super-tough substance, growing a garden, saving the planet, or just living your life: When you follow God's methods, patterning your garden, your life, or whatever you do after His ways, in time success will follow. He *"...is able to do exceeding abundantly above all that we ask or think, according to the power that worketh in us"* (Ephesians 3:20).

NOTES ON CHAPTER 7

1. Philip Hoare, "Super-Strong Limpet Teeth: Let's Hang on to Their Place in Nature," *The Guardian* web site, February 18, 2015, para. 2.
2. David Gutierrez, "Mother Nature's Nanotechnology: Oak Leaf Turned into Functioning Battery," *NaturalNews* web site, March 5, 2016, http://www.naturalnews.com/053195_oak_leaf_sodium_battery_Mother_Nature.html, para. 2.
3. A redwood named Hyperion has been measured at a truly amazing 379.7 ft. "Giant Sequoias and Redwoods: The Largest and Tallest Trees," LiveScience web site, https:// www. livescience.com/39461-sequoias-redwood-trees.html, para. 12.
4. Felicia Fonseca, "Feds Expand Efforts to Fight Wildfires by Reshaping Forests," AP News web site, May 4, 2015, https:// www. apnews. com/1bc-9f6a46dfa4cde84ce f83fe4043b3c, paras. 2–3, 10, 13.
5. Paul Gautschi, Back to Eden [Film], 2011, 00:05.
6. Melodie Metje, "Weed-Free, Self-Fertilizing, Till-Free Garden Beds," Victory Garden on the Golf Course [Blog], May 13, 2017, http://victorygardenonthegolfcourse.blogspot.com/2014/10/weed-free-self-fertilizing-till-free.html.
7. Tita Horta e Costa, "'Life in Syntropy'—Ernst Gotsch in Lisbon," A Sustainable Living [Blog] http://sustainportugal.eco/2017/10/03/life-in-syntropy-ernst-gotsch-in-lisbon/
8. "Mount St. Helens 30 Years Later: A Landscape Reconfigured," *Science Update* 19, Spring 2010, https://www.fs.fed.us/pnw/pubs/science-update-19.pdf.
9. In March 1980 the Georgia Guidestones (called the American Stonehenge) located in Elbert County, Georgia, U.S.A. were unveiled to the public. This strange monument consists of five massive 16 ft tall slabs (212,746 lbs) of polished granite, and one 25,000-pound capstone that rests upon the five. One slab stands in the center of the group, with the other four arranged around it; all are astronomically aligned. Extraordinary measures were taken to ensure that the group which financed the monument was kept secret and anonymous forever. Thus, no one knows who commissioned it or why. There is also a lot of speculation surrounding its meaning, for the designer was never named, and the reason was never stated. Inscribed on the structure in eight modern languages (English, Spanish, Swahili, Hindi, Hebrew, Arabic, Chinese, and Russian) is a set of 10 guidelines.

   1. Maintain humanity under 500,000,000 in perpetual balance with nature.
   2. Guide reproduction wisely—improving fitness and diversity.
   3. Unite humanity with a living new language.
   4. Rule passion—faith—tradition—and all things with tempered reason.

5. Protect people and nations with fair laws and just courts.
6. Let all nations rule internally resolving external disputes in a world court.
7. Avoid petty laws and useless officials.
8. Balance personal rights with social duties.
9. Prize truth—beauty—love—seeking harmony with the infinite.
10. Be not a cancer on the earth—Leave room for nature—Leave room for nature.

This *new world order/new age* monument is very chilling and unsettling when you contemplate its implications, especially in light of the first guideline—genocide on a massive scale. As of February 2018 the current world population is 7.6 billion. This means 7.1 billion people will have to be eliminated, and to maintain their ideal number of inhabitants at less than half a billion, strict measures and controls would be enforced.

10. Louise Gray, "David Attenborough—Humans Are Plague on Earth," *The Telegraph* web site, January 22, 2013, http://www.telegraph.co.uk/news/earth/earthnews/9815862/Humans-are-plague-on-Earth-Attenborough.html, paras. 1–4.
11. A. Kent MacDougall, "Humans as Cancer," Church of Euthanasia web site, n.d., http://www.churchofeuthanasia.org/e-sermons/humcan.html.
12. Matt Cardin, "It's Official: The Human Race is Earth's Disease," The Teeming Brain web site, May 9, 2009, http://www.teemingbrain.com/ 2009/ 05/09/its-official-the-human-race-is-earths-disease/.
13. Agent Smith Monologue at 1:37:35, The Matrix, dir. Lana Wachowski, Lilly Wachowski, (1999; Burbank, CA: Warner Bros. Pictures).
14. Ibid. para. 12.

In 1973, I proposed that our Universe had been created spontaneously from nothing (*ex nihilo*), as a result of established principles of physics. This proposal variously struck people as preposterous, enchanting, or both.

The novelty of a scientific theory of creation *ex nihilo* is readily apparent, for science has long taught us that one cannot make something from nothing.*

CHAPTER 8

BALONEY ON THE MENU

Growing up, I would always dread the answer to my question "what's for supper tonight?", fearing that my mother would reply "baloney." Every once in a while, dad would bring home a large ring of baloney from a downtown butcher shop.[1] Mom would dutifully boil it in a hefty kettle, deposit it on a large platter, and place it on the table. Dad would then slice off a good portion and plop it down on a plate for each member of our family.

I despised the taste of baloney and its kindred sister the hotdog, but no matter how much I complained, my parents expected me to eat what was on my plate, and that meant the whole hunk of "meat" that was now nestled between the creamed corn, applesauce, and mashed potatoes. To dispose of the meat substance as quickly as possible, I would cut it into small pieces and stuff the bits into my mouth without swallowing. I hoped that when I did so, I did not look like a chipmunk with its checks puffed-out with nuts. Trying not to draw attention, I

---

*Edward P. Tryon (Professor of Physics, City University of New York, USA), 'What made the world?' *New Scientist*, 8 March 1984, p. 14.

would than get up slowly from the table, walk calmly to the bathroom, and deposit the contents packed tightly into my mouth into the waiting toilet bowl to be flushed away. I thought it would be a good food source for the mutant alligators I was told lived in the sewers under our streets.

I have eaten pork and beans, pork chops, pork sausage, roast pig, bacon, and fried spam. It was easy to ascertain the source of those products, for the meat of a pig has that certain flavor that, it seemed to me, would be difficult to disguise even when mixed in with other ingredients. I liked the taste of the hog, so because baloney never had the slightest hint of pork, and tasted so awful, I was sure that none of its ingredients came from a pig.

Have you ever wondered what kind of animal is made into the meat we call baloney? For you will never see a herd of baloney grazing on the grass, hear a flock of those critters quacking as they fly overhead on their way to another pond, nor witness a school of baloney showing off their synchronized swimming abilities. So if you like the taste of baloney and would like to continue enjoying it at mealtime, do not ponder that question too deeply for any length of time.

Of course, the principal usage of the word baloney is to identify something that is nonsense, hogwash, or gibberish, which also happens to be the defining characteristic and accurate description of the belief in the *religion of evolutionism*. I think most people would hold in suspicion, cast a wary eye, and have great misgiving of a supposedly educated adult who truly believed in Santa Claus. I would have great apprehension of the sanity and lucidity of someone who believed that while a child slept, the tooth fairy makes her rounds to exchange lost teeth for money. And in the same fashion, I have great doubt regarding someone's rationality, logic, and scientific literacy who believes that given enough time, rocks will evolve into people.

I have great skepticism, reservation, and mistrust (and rightly so) for any so-called science article that comes from those who espouse or hold to the teaching of evolutionism. Their irrational and groundless beliefs thoroughly permeate the article and mix any science it might contain with ramblings of imaginary, impossible events, so much so that it becomes very hard to distinguish genuine facts from conjecture, science from inference, and scientific assumptions from thinly disguised religious ideology.

Evolutionary delusions have infiltrated every scientific discipline and have inexorably wormed their way into the thinking, rationale, and assessments of many teachers and scientists. Their adherence to this abhorrent religion limits their respective specialties' effectiveness, value, and benefit to mankind. Spending any money to determine how reindeer fly (how evolution happened), is a vain attempt to fill up a bottomless pit of lies. Just as baloney is a mixture of meat scraps, evolutionism is an assortment of make-believe, fantasy, and nonsense that is flavored and thoroughly homogenized with an extra measure of surreal religion.

The well-known scientist Dr. Henry M. Morris said it well:

> The faith of the evolutionist . . . is a splendid faith indeed, a faith not dependent on anything so mundane as evidence or logic, but rather a faith strong in its childlike trust, relying wholly on omniscient Chance and omnipotent Matter to produce the complex systems and mighty energies of the universe. The evolutionist's faith is not dependent on evidence, but is pure faith—absolute credulity.[2]

Professor of philosophy and zoology at the University of Guelph, Dr. Michael Ruse, defined evolution's ideology as follows:

> Evolution is promoted by its practitioners as more than mere science. Evolution is promulgated as an ideology, a secular religion—a full-fledged alternative to Christianity, with meaning and morality. I am an ardent evolutionist and an ex-Christian, but I must admit that in this one complaint— and Mr. Gish is but one of many to make it—the literalists

are absolutely right. Evolution is a religion. This was true of evolution in the beginning, and is true of evolution still today.[3]

Sir Arthur Keith wrote: "Evolution is unproved and unprovable. We believe it only because the only alternative is special creation, and that is unthinkable."[4]

Professor Louis Bounoure was Director of the Strasbourg Zoological Museum. In the March 8, 1984 edition of the *French National Centre Advocate*, he had this to say about evolution: "Evolution is a fairy tale for grownups. The theory has helped nothing in the progress of science. It is useless."[5] Journalist and satirist Malcolm Muggeridge wrote:

> I myself am convinced that the theory of evolution, especially the extent to which it has been applied, will be one of the great jokes in the history book of the future. Posterity will marvel that so flimsy and dubious an hypothesis could be accepted with the incredible credulity that it has.[6]

Dr. T. N. Tahmisian, physiologist for the Atomic Energy Commission, had this to say about evolution:

> Scientists who go about teaching that evolution is a fact of life are great con men, and the story they are telling may be the greatest hoax ever. In explaining evolution we do not have one iota of fact. A tangled mishmash of guessing games and figure juggling [Tahmisian called it].[7]

It seems that in the institutions of higher education, such as colleges and universities, you will often find a greater percentage of instructors who profess a belief in evolutionism. Professor James J. Krupa is one such professor. He expounds human evolutionism at the University of Kentucky, an institution steeped in a history of defending evolution education. He is what some would term a militant, rabid, intense, and bigoted evangelist. He is an enthusiastic advocate, an exemplar, and ultimate representation of the term *religious zealot*. In the March/April 2015 issue of *Orion*, he wrote an essay titled *Defending Darwin*. At the beginning of his essay he lamented, "There are some students I'll never reach."

A teacher does not defend science, data, statistics, or facts; he or she just relays the information to the student. Teachers do not reach students in order to impart a faith that the Mississippi River flows southward into the Gulf of Mexico, that there are planets in our solar system, that George Washington was the first president of the United States, or that American inventor Eli Whitney invented the cotton gin in 1793. Those are facts and histories that educators teach, hopefully in a stimulating manner. Schoolteachers need to provide stimulus and inspiration, a purpose, and an incentive to spur their students to study and do their best. Most importantly, a good teacher will teach students how to think, not what to think.

The "reaching" Krupa referred to has a religious connotation. He wants to evangelize his students, for he is a faithful and staunch devotee of the "Darwinian gospel of death." He wants to change his students' belief system, not impart new scientific information, facts, data, or knowledge. He wants to convert, persuade, proselytize, and bring them around to faith in his religion: evolutionism. In other words, he wants to give them a full dosage of baloney:

> I realized early on that many instructors teach introductory biology classes incorrectly. *Too often evolution is the last section to be taught*, an autonomous unit at the end of the semester. I quickly came to the conclusion that, *since evolution is the foundation upon which all biology rests, it should be taught at the beginning of a course, and as a recurring theme throughout the semester*. My basic biology for nonmajors *became evolution for nonmajors … evolution is the foundation of our science …* as a biologist, *the mission of advancing evolution education is the most important aspect of my job … to shy away from emphasizing evolutionary biology is to fail as a biology teacher*.[8] (emphasis added)

As you can read in his own words, Professor Krupa is certainly an opinionated, prejudiced, and narrow-minded ideologue. I am sorry to say he is not just an isolated individual; he is part of an evil, well-entrenched atheistic cartel of like-minded individuals.

In Professor Krupa's essay, he makes a startlingly contradictory statement, for if someone truly understands science and allows its methods to rule all his findings and conclusions, there would automatically follow a rejection and revulsion of evolutions' ideology:

> *To truly understand evolution, you must first understand science*... The National Academy of Sciences provides concise definitions of these critical words: A *fact* is a scientific explanation that has been tested and confirmed so many times that there is no longer a compelling reason to keep testing it; a *theory* is a comprehensive explanation of some aspect of nature that is supported by a vast body of evidence generating *testable* and *falsifiable* predictions. (emphasis added)[8]

As evolutionism is a belief about the history of how things happened in the distant past, in just what way could its study be called science? After all, you cannot observe or test history, two criteria that are absolutely necessary for scientific investigation. You also have to be able to make falsifiable predictions, but that cannot be done when you are dealing with history. When I say evolution cannot be falsified, I mean no tests can be performed that would verify that it happened.

DON'T GET TAKEN FOR A RIDE WITH NASA

On October 1, 1958, a new agency became operational that replaced the National Advisory Committee for Aeronautics (NACA): The National Aeronautics and Space Administration (NASA). NASA is a U.S. government agency with a distinctly civilian (rather than military) orientation that encourages peaceful applications of technology dealing with airplanes and space science. Now, while the intentions of those whose efforts and labors helped birth this government agency may have been benign, I have a few problems with a government agency that is involved with things I believe may conflict with the U.S. Constitution and should be in the hands of private industry, but whether I am right or wrong on that matter is another issue altogether. The real problem I have with tax dollars being spent in this manner is the great amount of money being wasted, which is in effect being spent by those trying to find out how Santa's reindeer fly. Yes, baloney is on their menu.

For example, on the NASA Astrobiology Institute's official site, a post titled "About Astrobiology" was published on December 7, 2014. It said:

> Astrobiology is the study of the *origin, evolution*, distribution, and future of life in the universe. This multidisciplinary field encompasses... [a] search for evidence of *prebiotic chemistry* and life on Mars and other bodies in our Solar System, laboratory and field research into the *origins and early evolution* of life on Earth... NASA's Astrobiology Program addresses three fundamental questions: *How does life begin and evolve? Is there life beyond Earth and, if so, how can we detect it? What is the future of life on Earth and in the universe?* In striving to answer these questions... experts in astronomy and astrophysics, Earth and planetary sciences, microbiology and *evolutionary biology*... and other relevant disciplines are participating in astrobiology research and helping to advance the enterprise of space exploration...
>
> Inner solar system bodies... *are thought to have formed from the accretion of dust into "planet-esimals," the planetesimals into proto-planets, and finally the proto-planets into planets.* Many details of this sequence are still unknown... although Venus, the Earth, and Mars are similar to one another, *they have evolved in highly distinct ways*... Missions currently in operation... will greatly increase our knowledge of the *forces driving planetary... evolution*... The giant

> planets...hold many *clues to the origin and evolution of our solar system*...(emphasis added)[9]

NASA's Astrobiology program addresses three fundamental questions here. If each area of research receives the same amount of money, one third of all funding would be going down a rabbit hole, for one of their three fundamental questions is, *How does life begin and evolve?* Life never just begins; that is a scientifically impossible event. Nor does it evolve as evolutionism instructs. The study of prebiotic chemistry is a part of that fundamental question, so what is prebiotic chemistry? Let us go to the Answers.com website for the answer. Under the heading *In Chemistry* is this question: "What is prebiotic chemistry?"

> Prebiotic chemistry is the field of study involving the spontaneous chemical reactions which may have led to the formation of biomolecules and/or life on early Earth. In general, prebiotic chemistry includes all possible abiotic reaction pathways leading from inorganic substances to organic substances to biomolecules or bio-like molecules.[10]

Evolutionists' "prebiotic chemistry" experiments are not non-biased, for they start with preconceived notions, beliefs, and foregone conclusions: First, that life arose from inorganic matter; second, that they have a ballpark idea of the chemicals that were used to form the first simple prebiotic machines; and third, they have an understanding of the conditions that were present in the natural environment on the early Earth, about 4 billion years ago before the advent of life. Weather forecasts for the next day are often incorrect, but evolutionists know what the weather conditions were like four billion years ago.

Journalist Gregg Easterbrook has stated correctly the evolutionists' non-understanding of the origins of life: "What creates life out of the inanimate compounds that make up living things? No one knows. How were the first organisms assembled? Nature hasn't given us the slightest hint. If anything, the mystery has deepened over time."[11]

Evolutionists have attempted to test a few of their beliefs. So what did their tests prove? The most infamous prebiotic chemistry experiment in history was carried out by Miller and Urey at the University of Chicago in 1952. For all their effort and time, their experiment managed to create a flask of poisons, and was able to generate quite a bit of tar—a fine substance to pour into a crack in the road but not one that can form life. Simply put, *prebiotic chemistry is a way to determine how processes that never happened, under conditions which never existed, during a time that never was, made it possible to create an organism that never lived, using a synthesis of materials brought into being by nothing.* Its study should rightly be called baloney.

Education is the acquisition of true wisdom and true knowledge, so those who believe in the *religion of evolutionism* are not truly educated.

BALONEY DETECTION KIT

I have recently found a useful tool to help me move rationally through opinions, ideas, concepts, and philosophies found floating all around us every day. It's a series of questions and principles put forth and used by many scientists. It's called the *Baloney Detection Kit.*

What is interesting about the two kits I found on the Internet is that the questions are from two die-hard evolutionists, Richard Dawkins and Carl Sagan. The real shame is that they still miss the mark; they cannot see the error of their own beliefs even though they have put together a list of queries that, for someone who is thinking logically and rationally, should be enough to allow him or her to see the "light." The following information, from the article "How to Assess the Believability of Claims without Succumbing to Cynicism," comes from Maria Popova's interesting website, *Brain Pickings*:

> "Baloney Detection Kit" for grown-ups from the Richard Dawkins Foundation for Reason and Science and *Skeptic Magazine* editor Michael Shermer—a 10-point checklist for

assessing the believability of a claim, covering everything from telling the difference between science (e.g., SETI) and pseudoscience (e.g., UFOlogy) to detecting personal agendas.

The complete checklist:
1. How reliable is the source of the claim?
2. Does the source make similar claims?
3. Have the claims been verified by somebody else?
4. Does this fit with the way the world works?
5. Has anyone tried to disprove the claim?
6. Where does the preponderance of evidence point?
7. Is the claimant playing by the rules of science?
8. Is the claimant providing positive evidence?
9. Does the new theory account for as many phenomena as the old theory?
10. Are personal beliefs driving the claim?[12]

These are all great questions, but for the sake of brevity, I will only take the time to comment on numbers 4, 7, and 10.

Question number four asks, "Does this fit with the way the world works?" For evolution, as already pointed out in a number of places in this book, the answer is: Absolutely Not!

Number seven asks, "Is the claimant playing by the rules of science?" For the most part, beliefs concerning events that happened in the past are outside of the scope of scientific investigation, so what are the rules of science, otherwise known as the scientific method?

The scientific method usually has five or six basic steps that are taken before scientists can come to any kind of conclusion. Scientists will generally follow these steps when considering a problem. First they conduct observation, formulate a question, do background research, and then develop or construct a hypothesis (a testable prediction). They will then refine the idea and make predictions according to the hypothesis. They will devise experiments and tests (which must be repeatable) to confirm or prove their hypothesis, and then analyze the data from the tests and experiments. They generally try to falsify the hypotheses by the tests they devise, as the purpose of their experiments is to determine whether things observed in the real world agree or conflict with the predictions derived from a hypothesis. Lastly, they will draw conclusions from those experiments and make a final analysis and statement.

How do you test, measure, observe, or experiment on history? What kind of experiments or tests could you devise or design concerning history that would enable you to disprove your hypothesis? Any hypothesis put forward must be testable and potentially falsifiable, which means that there must be a way to disprove and show that it is false. However, a hypothesis can never truly be proven using the scientific method, for there are always other explanations that can explain the results, although they can be disproven, in which case the hypothesis is rejected as false. Any tests scientists devise or can conceive of must be repeatable and have measurable results with empirical (knowledge acquired by means of observation or experimentation) evidence, such as occurs in medical trials and various treatments. Given these facts, the scientific method cannot be used to prove evolutionary history or the events that happened during the past in the lifetime of any person or living creature.

And now for question number ten: "Are personal beliefs driving the claims?" The evolutionist's ideology and worldview most certainly drive all of their endeavors, activities, and conclusions.

On the November 9, 2014 blog titled Climate Etc., there was a post from Popova: "The Baloney Detection Kit: Carl Sagan's Rules for Bullshit-Busting and Critical Thinking." It states:

> In *The Demon-Haunted World: Science as a Candle in the Dark*, Sagan shares his secret to upholding the rites of reason, even in the face of society's most shameless untruths and outrageous propaganda.
>
> In a chapter titled "The Fine Art of Baloney Detection," Sagan reflects on the many types of deception to which we're susceptible—from psychics to religious zealotry...

> Through their training, scientists are equipped with what Sagan calls a "baloney detection kit"—a set of cognitive tools and techniques that fortify the mind against penetration by falsehoods. Sagan shares nine of these tools...[13]

Each one of Sagan's nine tools is an outstanding guide to help someone properly focus and channel their thinking and line of reasoning. However, we will not list all nine points; rather, we will spend a little time on his ninth principle:

> Always ask whether the hypothesis can be, at least in principle, falsified. Propositions that are untestable, unfalsifiable are not worth much. Inveterate skeptics must be given the chance to follow your reasoning, to duplicate your experiments and see if they get the same result.[14]

For upwards of 50 years, many high school chemistry labs have routinely repeated the 1953 Miller and Urey's classic fiasco to create life, each time with the same result: failure. I love Sagan's statement: "Propositions that are untestable, unfalsifiable are not worth much." So, as we learned in response to number seven of Dawkins' *Baloney Detection Kit* for grown-ups, evolution is outside of the scope of science; religious beliefs cannot be tested, which means they are most certainly unfalsifiable. Thank you for your insight, Mr. Sagan.

I agree completely with one of Judith Curry's comments on the web site Climate, Etc.: "These 'rules' are useful common-sense reminders for evaluating any sort of claim. Too often serious baloney detection is ignored by scientists in the interests of careerism and advocacy."[15] This comment certainly unveils the hallmark of those who espouse the teachings of evolutionism, for they fail to heed its advice.

NOTES ON CHAPTER 8

1. Bologna sausage is sometimes phonetically spelled baloney. It is a finely ground pork sausage that contains cubes of lard mixed with seasonings, such as black pepper, nutmeg, allspice, celery seed, myrtle berries, and coriander. It can also be made out of beef, turkey, chicken, pork, venison, soy protein, or almost any combination of "meat" scraps or parts.
2. H. Morris, *Some Call It Science*, revised ed. (Dallas, TX: Institute for Creation Research, 2006), p. 7.
3. Michael Ruse, "How Evolution Became a Religion," Omniology.com web site, May 13, 2000, www.omniology. com/ How Evolution Became Religion, para. 7.
4. https://www.secular-humanism.com/, para. 5.
5. Louis Bounoure, *Pascal Lectures*, printed in the *French National Centre Advocate*, March 8, 1984, p. 17, quoted by Donald Honeycutt, *The Origin Debate/ Who, or What is Responsible for This* (Bloomington, IN: Xlibris Corporation, 2010), p. 33.
6. Malcolm Muggeridge, quoted on the Goodreads web site, n.d., https://www.goodreads.com/quotes/913269-i-myself-am-convinced-that-the-theory-of-evolution-especially
7. T.N. Tahmisian, [article title unknown], *The Fresno Bee*, August 20, 1959, p. I-B, quoted in "Paleomagnetism," Evolution Encyclopedia vol. 3, God Rules web site, http://www.godrules.net/ evolutioncruncher/ 3 evlch26.htm, para. 1.
8. James Krupa, "Defending Darwin," Slate.com web site, March 26, 2015, http://www.slate.com/articles/health_and_science/science/2015/03/teaching_human_evolution_at_the_university_of_kentucky_there_are_some_students.html, para. 10.
9. NASA, "About Astrobiology," NASA Astrobiology Institute web site, December 7, 2014, https://nai.nasa.gov/about-astrobiology/, paras. 1–2.
10. http://www.answers.com/Q/What_is_prebiotic_chemistry.
11. Gregg Easterbrook, "Where Did Life Come From? *Wired*, February 2007, p. 108.
12. Maria Popova, "The Baloney Detection Kit: A 10-Point Checklist for Science Literacy," Brain Pickings Web site, March 16, 2012, https://www.brainpickings.org/2012/03/16/baloney-detection-kit/
13. Maria Popova, "The Baloney Detection Kit: Carl Sagan's Rules for Bullshit-Busting and Critical Thinking," Brain Pickings web site, January 3, 2014, https://www.brainpickings.org/2014/01/03/baloney-detection-kitcarl-sagan/, paras. 1–2.
14. Carl Sagan, quoted in Popova, 2014, para. 6.
15. Judith Curry, "Sagan's Baloney Detection Rules," Climate Etc. web site, November 9, 2014, https://judithcurry.com/2014/11/09/sagans-baloney-detection-rules/, para. 9.

> Contrary to what most scientists write, the fossil record does not support the Darwinian theory of evolution because it is the theory (there are several) which we use to interpret the fossil record. By doing so we are guilty of circular reasoning if we then say the fossil record supports this theory.*

CHAPTER 9

THERE'S A SUCKER BORN EVERY MINUTE

Phineas Taylor Barnum, better known as P. T. Barnum (July 5, 1810–April 7, 1891) is largely remembered for circus sideshows that displayed freaks, for promoting celebrated hoaxes, and for founding the Barnum and Bailey Circus. After a merger in 1919, it was known as the Ringling Bros. and Barnum & Bailey Circus, and billed as "The Greatest Show on Earth." (After entertaining an estimated 250 million people over the course of 146 years, the Greatest Show on Earth bowed out on May 21, 2017. After years of pressure from animal rights activists, in May 2016 the final performances with elephants took place. The decline in attendance due to the lack of the elephants was greater than anticipated, and that, together with increasing operating costs and lower ticket sales, forced the closure.) Barnum was also an author, publisher, politician (he was the mayor of Bridgeport, CT), and a philanthropist. During his lifetime, he considered himself a showman by profession. It has been said of him that his personal aim was "to

---
*Ronald R. West, Ph.D. (paleoecology and geology) (Assistant Professor of Paleobiology at Kansas State University), 'Paleoecology and uniformitarianism'. *Compass*, vol. 45, May 1968, p. 216.

put money in his own coffers." At his death, he was perhaps the most famous American in the world, so much so that even his detractors and critics hailed him as an icon of the American spirit and ingenuity. Barnum is commonly credited with coining the phrase, "There's a sucker born every minute."

Unfortunately, his biographer, Arthur H. Saxon, was unable to verify that Barnum ever uttered this phrase. According to Saxon, "There's no contemporary account of it, or even any suggestion that the word 'sucker' was used in the derogatory sense in his day. Barnum was just not the type to disparage his patrons." The comment was most likely uttered by David Hannum, a competitor of Barnum, who spoke it in reference to Barnum's part in the Cardiff Giant hoax,[1] the most talked about exhibit in the nation at the time:

> Hannum, who was exhibiting the "original" stone giant and had unsuccessfully sued Barnum for exhibiting a copy and claiming it was the original, was referring to the crowds continuing to pay to see Barnum's exhibit even after both it and the original had been proven to be a fake.[2]

While Hannum's name was lost to history, Barnum was left with the misplaced stigma for being the man who said "there's a sucker born every minute."[3] One fascinating fact about this story is not that the public was totally taken in; at first neither exhibitor Barnum or Hannum was aware that Hannum's stone giant was a fake. The only people not taken in by this hoax were those who had carved the original stone giant and the perpetrator of the hoax, Mr. George Hull, of Binghamton, NY.

Read a little more and see if you have been or are now a sucker for a few of the deceptions, frauds, and hoaxes that infect the world today.

DEADLY HYDROHYDROXIC ACID

As I was doing research for this chapter, I happened upon some information about the growing use of a dangerous chemical compound, but what I found most alarming was that its utilization was not just limited to a few distant inaccessible mountain locations; I found it to be widespread across the general population of the Philippines. I wondered what the government might be doing to help combat this blight as it spread island to island, so I turned to the Philippine Law and Jurisprudence Databank for information on the law that was approved on March 30, 1972, known as "The Dangerous Drugs Act of 1972."

Reading through this regulation, I came to the section I was looking for: "Article II, Prohibited Drugs." I wondered if I would find within this law the action necessary to ban this chemical compound. There are various names for this element; among them is Hydrogen oxide, Dihydrogen monoxide (DHMO), and Hydrohydroxic acid. Search as I might, however, the element did not seem to be itemized among the many dangerous drugs listed.

This substance is colorless, odorless, tasteless, and can be found in three states—vapor, liquid, and solid. It is certainly dangerous in all its forms, for it kills uncounted thousands of people every year. Most fatalities occur when it is in its liquid state, caused by the accidental inhalation of this material. It is used as an industrial solvent and coolant; a fire-retardant material; it may cause severe burns, and is toxic in large quantities. It is a major component of acid rain and it contributes to the greenhouse effect. It is used in the growing of crops, but even after vigorous washing, produce remains contaminated by this chemical. It also seemed to have infiltrated much of the water purification infrastructure of our cities, for it has been found coming out of local water taps. And for those who have become dependent upon its use, withdrawal means certain death.

Besides these physical dangers, Dihydrogen monoxide has been found to contribute to the erosion of our natural landscape. It accelerates corrosion and rusting of many metals. It may cause electrical failures and decreased effectiveness of automobile brakes. The pollution from this chemical contaminant is global, for this substance has been found in almost every river, lake and reservoir in all known countries, has been found under the Antarctic ice sheet and even in breast milk. Because DHMO is still legal, companies dump this waste into rivers and the ocean where its effect on wildlife is extreme. We cannot afford to ignore these facts any longer; its contamination is reaching epidemic proportions! As a concerned environmentalist, I believe we must work to reduce the availability of this substance; its use must be restricted. First we should petition the Philippine government to amend "The Dangerous Drugs Act of 1972" to include DHMO. We must then work together to ban it as soon as possible!

As you ponder how this situation got so far out of hand, let me relay a little more information about this chemical. The better informed you are about what it is and its negative effects, the greater good you can do in the fight against it. The following details will help fill in the rest of the story and help you understand just what actions you may need to take to fight this great evil that has crept up on us all.

The substance known as Dihydrogen monoxide consists of two hydrogen atoms and one oxygen atom, which, when combined forms a molecule often referred to as H_2O, and is more commonly called *water*. You know...water...the stuff that falls from the clouds as rain or snow, the liquid we drink, bathe, cook, and swim in.

When you started to read this information, were you scientifically literate enough to catch this right away? This hoax is not original to me; it has been around for over twenty years and there are many versions of it. Just do an online search for information about Dihydrogen monoxide and a long list of articles and stories will come up.

As we read about the various ways and means this hoax has been presented, we find that people from all walks of life and professions, such as those in the Australian Parliament, city council members in Aliso Viejo, CA, Finnish parliament candidates, and people on the street have fallen for it and joined in efforts to have this "toxic substance," DHMO, banned.[4]

Wikipedia had this information concerning the Dihydrogen monoxide hoax:

In 2006, in Louisville, Kentucky, David Karem, executive director of the Waterfront Development Corporation, a public body that operates Waterfront Park, wished to deter bathers from using a large public fountain. "Counting on a lack of understanding about water's chemical makeup," he arranged for signs reading: "DANGER!—WATER CONTAINS HIGH LEVELS OF HYDROGEN—KEEP OUT" to be posted on the fountain at public expense.[5]

This hoax gained renewed popularity and wide-spread public attention in 1997 when a 14-year-old junior high student from Idaho Falls, ID, Nathan Zohner gathered petitions to ban DHMO as the basis of his science project about gullibility, aptly titled "How Gullible Are We?" This story has since been used in science classes to encourage critical thinking. The list about Dihydrogen monoxide is long and scary and contains absolutely true warnings about it. It is used in the production of Styrofoam, the nuclear power industry, is present in cancer tumors, and is guaranteed fatal to most animals and people in large quantities. People tend to assume that something this scary is bad for you, even though it is necessary for life.

The whole point of this was to show how easy it is to fool some people into thinking something is scary when it is stated in an alarming manner. Using misleading language and statements, using facts, correct information, statistics, and data, one can create a false argument or conclusion that will mislead the general public. As far as I am aware, this hoax has not caused physical harm, damage or injury to anyone, though it has damaged some folks' pride.

The hoax of evolutionism, however, has caused untold harm that really has led many astray. It has created scientific-sounding terms to add weight and acclaim to its exaggerated analysis. Along with a deep amount of gullibility, there is a general lack of scientific literacy among the general public. There is also a disregard of logic, observation, and some scientific procedures by all scientists who teach this nonsense.

Many have been persuaded to trust and believe statements and facts that are actually false, misleading, and meaningless.

We will deal a little more with that philosophy shortly, but for this next section, the word "hoax" might not be the best word to use; deception, conspiracy, and cunningly conceived plot describe more fully what has been perpetrated on the United States, the Philippines, and most other nations on this planet.

GOD'S BENEFACTION TO MANKIND

Much of the following information has been left out of our education. Much of this knowledge, evidence, and information have been covered up and censored. The remaining crumbs fed to us are largely slanted and distorted by legislators, government officials, and those in positions of authority. We often believe lies told to us constantly by those in places of trust and responsibility without any evidence, verification, support, or proof. When there is official disapproval and censure by legislative votes, consistent public statements by officials, constant reports from television newscasts detailing dangers, and repetitious stories in newspapers and magazines condemning and denouncing something for decades, the majority of the public's sentiment, right or wrong, will most often side with such pronouncements.

Due to misrepresentation and propaganda, I was very skeptical at first (and many still are), but I found the evidence to be so overwhelming that I had no choice but to set aside what those in power were saying and act accordingly. It took research and open-mindedness for me to recognize the falsity of the supposed truths I was being told. I hope the reader will be able to put aside biased, prejudiced, and predisposed beliefs while considering the following material.

A wise man once said that when you are considering the veracity of something, "Follow the money trail." See who has something to gain or lose if the thing being considered is prohibited, if permission is denied, or if it is allowed to operate or to be produced. While this book is concerned with the fraudulent teaching of evolutionism as science, this next portion, although it does not touch upon that subject in particular, is pertinent to our discussion, for it deals with and helps us understand the way the public can be misled, misinformed, taken in, and greatly harmed by so-called information coming from science via the media and those who are in positions of hegemony and authority.

This next scheme and ruse has been cunningly planned, laid out, and executed by malicious men in places of power and influence. I believe that true Christian faith encompasses and holds to a worldview that is concerned for everything that takes place on Earth. God asked Job a rhetorical question: "*Who provideth for the raven his food?*" (Job 38:41). The answer is the Creator. Jesus instructed the multitudes on the mount about God's loving compassion and concern for people and His creation: "*Behold the fowls of the air: for they sow not, neither do they reap, nor gather into barns; yet your heavenly Father feedeth them*" (Matthew 6:26).

Paul, writing to the Church in Philippi, instructed them: "*Let each of you look out not only for his own interests, but also for the interests of others*" (Philippians 2:4 NKJV). Therefore, as a Christian, I am concerned with the economic, financial, commercial, trade, and industry in this poor nation, the Philippines, which I now call home. I am personally aware of fathers who struggle and work hard to provide for their families, and yet their children do not have enough to eat at times and go to bed hungry. I know mothers and fathers who have had to work abroad for years to have money to send back to their families so their children could attend school. So when I see the prohibition by the government of this nation and most of the world concerning the growing and control of a certain plant, I want to expose every evil way men wage war, their use of guile, malice and hidden craftiness so the lives of people here and all over the world will be enriched.

The growing of this plant, with its multitude of uses, would be a godsend to the economy of this poor nation and a welcome means of good income for many. Out of the entire collection of flora, foliage, and shrubbery, I believe God's greatest gift to mankind is one plant in particular. Out of all the vegetation in existence, this plant contains the most powerful healing agents ever known, and coupled together with its numerous other benefits and uses, make it the most significant, valuable, noteworthy, versatile, and beneficial. For many centuries, most of the riggings and other necessary parts and components to wooden ships were made from this plant.[6] Great artists, such as Rembrandt, Gainsborough, Van Gogh, and others, painted primarily on canvas made from this plant. It makes superior fiber and paper products. Food, oils, proteins, and over 25,000 textiles and fabrics can be made from this plant. It is also used to make many types of building and construction materials. Composite building materials made from this plant have superior strength, flexibility, and economy when compared to wood fiber, even as beams. Strong, inexpensive, fire-resistant paneling with excellent thermal and sound-insulating qualities can be made and used in place of drywall and plywood.

From the *Hidden Mysteries: Conspiracy Archive* site comes this interesting information about Henry Ford using this plant along with other materials to produce his biological car:

> On August 14, 1941, at the 15th Annual Dearborn Michigan Homecoming Day celebration, Henry Ford unveiled his biological car. Seventy percent of the body of the cream-colored automobile consisted of a mat of long and short fibers from field straw, cotton linters, hemp, flax, ramie and slash pine. The other 30 percent consisted of a filler of soymeal and a liquid bio resin.
>
> The timing gears, horn buttons, gearshift knobs, door handles and accelerator pedals were derived from soybeans. The tires were made from goldenrods bred by Ford's close friend Thomas Edison. The gas tank contained a blend: about 85 percent gasoline and about 15 percent corn-derived ethanol.
>
> To prove the vehicle's superiority, Ford demonstrated the strength of the car body by smashing an axe against the trunk, only to have it bounce off.[7]

Its fiber is the strongest, most durable, longest-lasting overall natural soft-fiber on the planet. It's softer than cotton, warmer than cotton, more water absorbent, more air-permeable, has eight times the tensile strength and four times the durability. Because of these qualities, it continued to be the second most-used natural fiber until the 1930s. Unlike cotton, this plant has a small environmental impact. It is environmentally friendly to grow; it requires no harmful chemicals, no pesticides, no herbicides, little or no fertilizer, and has few weed, insect, or pest enemies. Growing cotton, on the other hand, has a heavy detrimental effect on the environment; approximately 50 percent of all chemicals used in American agriculture today are used in the production of cotton.

By using different strains of this plant, it can be grown in virtually any climate or soil condition on Earth. This plant has been used by many cultures from the beginning of recorded history, has literally tens of thousands of uses, produces seeds unsurpassed in nutritional value, and produces medicines that can cure cancer and other diseases.

Because of the numerous medical benefits and healing qualities of this plant, it was the first to the third most-important and most used medicine for two-thirds of the world's people for at least 3,000 years. Until the turn of the twentieth century, its extracts, made from the buds, leaves, roots, etc., were the most commonly used and widely accepted medicines in the world for the majority of human illnesses. Recent research into this plant has demonstrated therapeutic value and complete safety in treating many health problems, including Alzheimer's disease, anorexia, asthma, arthritis, depression, epilepsy, glaucoma, herpes, infection, migraines, multiple sclerosis, nausea, rheumatism, stress, and tumors.

Throughout the 1800s and until the 1930s, various extract medicines for human and veterinary uses were produced by apothecaries and many American and European companies, such as Eli Lilly, Smith Brothers, and Squibb. But what is unparalleled by any other kind of drug, medicine, or supplement modern drug companies produce, is that during all that time, there was not one reported death and virtually no abuse or mental disorders reported from medicines made from this plant's extract—except for first-time or novice-users occasionally becoming disoriented or overly introverted.

Before I disclose the name of this plant, let me provide a little more historical information. During the war for U.S. independence, the "Daughters of the American Revolution" organized spinning bees to help clothe George Washington's soldiers. The majority of the thread was spun from this plant. This plant was a staple crop of the early colonies. You could pay your taxes with this plant throughout America for over 200 years, for it was legal tender in most of the Americas from 1631 until the early 1800s. U.S. presidents George Washington, John Adams, Thomas Jefferson, and others grew this plant on their plantations. For the U.S. census of 1850 they counted 8,327 plantations. These were farms with a minimum of 2,000-acres that grew this plant for canvas, cloth and cordage used for baling cotton. This census did not include the tens of thousands of smaller farms, nor the hundreds of thousands of family garden patches in America that grew this plant.

Jack Herer, in his book *The Emperor Wears No Clothes*, says:

> Until the 1880s in America (and until the 20th century in most of the rest of the world), 80 percent of all textiles and fabrics used for clothing, tents, bed sheets and linens, rugs, drapes, quilts, towels, diapers, etc., and even our flag, "Old Glory," were principally made from fibers of [H].
>
> From more than 1,000 years before the time of Christ until 1883 A.D... was our planet's largest agricultural crop and most important industry, involving thousands of products and enterprises; producing the overall majority of Earth's fiber, fabric, lighting oil, paper, incense and medicines. In addition, it was a primary source of essential food, oil, and protein for humans and animals.
>
> [H] can be pressed for its highly nutritious vegetable oil, which contains the highest amount of essential fatty acids in the

plant kingdom ... seed protein is one of mankind's finest, most complete and available-to-the-body vegetable proteins ... is the most complete single food source for human nutrition.[8]

It has been said that this plant is many times more versatile than the cotton plant, the Douglas fir, and the soybean combined. What is the name of this miracle plant, earth's foremost renewable natural resource, God's benefaction, His endowment, and His bequest given freely to mankind? HEMP! It is also known as cannabis hemp, India hemp, Manila hemp, pot, marijuana, reefer, grass, ganja, etc. All are names for the same plant. Although hemp and marijuana are the same kind of plant, there are a number of strains and varieties that are grown for different reasons and purposes. The strain of this plant known as hemp is used for paper, ropes, clothing, building materials and numerous other products. The other variety that has people "up-in-arms," is the kind that can be smoked, but those varieties also have numerous other uses in medicine and food. But unlike other supplements and drugs, for the most part, will not cause any lasting or permanent damage, nor will its use alone, kill you. Concerning prescription drugs, Lance D. Johnson states:

> ...for the most part, modern pharmacopeia is a powerful, dangerous drug cult masquerading as helpful medicine. The statistics do not lie. About 330,000 American and European patients die annually from appropriately prescribed prescription drugs. This is the fourth leading cause of death—iatrogenic deaths, or death by medicine.[9]

Depending on which sources you use, and what is included, the following statistics may seem to be bloated or underreported. In either case, tobacco, alcohol, legal and illegal drugs cause a multitude of deaths per year. Notice, however, this wonder plant—hemp—does not even contribute to this list. In the entire history of the human race, very few deaths can be attributed to this plant!

U.S. deaths per year:

> Tobacco: 480,000–556,000
> Alcohol: 88,000
> Prescription drugs: 100,000
> Aspirin: 20,000
> Cocaine and Heroin: 19,766
> Synthetic opioids fentanyl and tramadol: 9,580
> Methadone: 3,285
> Benzodiazepine: 8,791
> Hemp: 0[10]

If worldwide statistics were considered, there would be a staggering total number of drug fatalities and drug caused problems from drugs other than hemp. For the most part, this plant is not harmful to the mind or human body. It is not addicting, but like many other things, it can become habitual with long-term use. Typically, its users need never to spend time in addiction or rehab programs.

However, to make this a balanced review, the 2005 report[11] did list marijuana as a possible contributing factor in 279 deaths. But again, most of those deaths were due to the abuse of other substances along with the use of marijuana. In Joseph Cariz's August 16, 2016 article "Is it True that Marijuana Hasn't Caused Any Deaths, While Prescription Drugs Have Caused 100,000?, he stated:

> …the U.S. Drug Enforcement Administration drug sheet for marijuana reports that no deaths from marijuana overdose have ever been recorded.
>
> However, marijuana has played an indirect role in fatalities. Experts that we talked with agreed that the drug itself doesn't cause major acute health problems and is far safer than other medications. However, they said that it can still dangerously inhibit someone's ability to make safe decisions. […]

"Too often individuals cite that individuals haven't died from cannabis—I don't think that's true. It certainly can be argued that cannabis use has contributed to the deaths of individuals, such as due to impairment during driving," [Dr. Ryan Vandrey] said.

Dr. Jerome Avorn, a professor of medicine at Harvard Medical School, told us, "The main risk from marijuana is from the risky or stupid things people do after using it, such as driving, rather than from any toxic effects of the substance itself, which is remarkably safe." [. . .]

In summary, it appears that marijuana doesn't directly cause overdose deaths, but there are documented cases where it likely led to accidental fatalities. The exact number of marijuana-related deaths that occur annually is difficult to pin down.[12]

To sum this up, if hemp is used properly, there is very little danger of harm or death.

Believe it or not, before it was made illegal, there were high hopes placed upon the farming of this plant back in the 1930s. In February of 1938, *Popular Mechanics* printed an article titled "New Billion-Dollar Crop."[13] This was the first time a cash crop had a business potential that exceeded a billion dollars.

A year earlier, *Mechanical Engineering* had published a story titled "The Most Profitable and Desirable Crop that Can be Grown."[14] It stated that if hemp was cultivated using twentieth-century technology, it would be the single largest agricultural crop in the U.S. and the rest of the world. These stories were not about growing pot for smoking or recreational use. They were referring to cannabis hemp, which is grown for industry and numerous other purposes.

Jack Herer has stated, "In case you're wondering, there is no THC or 'high' in hemp fiber. That's right; you can't smoke your shirt! In fact, attempting to smoke hemp fabric, or any fabric, for that matter, could be fatal!"[15] Doug Yurchey, in his article "The Marijuana Conspiracy:

The Real Reason Why Hemp is Illegal," said:

> The truth is if marijuana was utilized for its vast array of commercial products, it would create an industrial atomic bomb! Entrepreneurs have not been educated on the product potential of pot. The super-rich have conspired to spread misinformation about an extremely versatile plant that, if used properly, would ruin their companies. (emphasis added)[16]

In 1942 the United States Department of Agriculture (USDA) produced a 14-minute film titled *Hemp for Victory*. This informational film prompted patriotic American farmers to grow 350,000 acres of hemp each year for the war effort. The film details the numerous uses of hemp, including cordage and cloth, as well as a detailed history of the plant's use. Until 1989, the U.S. government denied (lied) that it had produced this film or that such a film even existed, but thanks to the efforts of researchers like Jack Herer, a copy of the original film was found in the Library of Congress and the film can now be viewed online in its entirety. The following information comes directly from that film:

> [When] Grecian temples were new, hemp was already old in the service of mankind. For thousands of years, even then, this plant had been grown for cordage and cloth in China and elsewhere in the East. For centuries prior to about 1850, all the ships that sailed the western seas were rigged with hempen rope and sails. For the sailor, no less than the hangman, hemp was indispensable.
>
> Now with Philippine and East Indian sources of hemp in the hands of the Japanese...American hemp must meet the needs of our Army and Navy as well as of our industries...[to supplement] the Navy's rapidly dwindling reserves. When that is gone, American hemp will go on duty again; hemp for mooring ships; hemp for tow lines; hemp for tackle and gear; hemp for countless naval uses both on ship and shore. Just as

in the days when Old Ironsides sailed the seas victorious with her hempen shrouds and hempen sails. Hemp for victory![17]

Hemp's safety has often been called into question by the ignorant and misinformed, so this official information is very revealing.

> 3. The most obvious concern when dealing with drug safety is the possibility of lethal effects. Can the drug cause death?
> 4. Nearly *all* medicines have toxic, *potentially lethal* effects. But *marijuana is not such a substance.* 4. There is *no record* in the extensive medical literature describing *proven, documented cannabis-induced fatality.* (emphasis added)
> 5. This is a remarkable statement. First, the record on marijuana encompasses 5,000 years of human experience. Second, marijuana is now used daily by enormous numbers of people throughout the world. Estimates suggest that from twenty million to fifty million Americans routinely, albeit illegally, smoke marijuana without the benefit of direct medical supervision. Yet, despite this long history of use and the extraordinarily high numbers of social smokers, there are simply no credible medical reports to suggest that consuming marijuana has caused a single death.
> 6. By contrast aspirin, a commonly used, over-the-counter medicine, causes hundreds of deaths each year...In order to induce death a marijuana smoker would have to consume 20,000 to 40,000 times as much marijuana as is contained in one marijuana cigarette...A smoker would theoretically have to consume nearly 1,500 pounds of marijuana within about fifteen minutes to induce a lethal response.
> 9. In practical terms, marijuana cannot induce a lethal response as a result of drug-related toxicity.[18]

Fresh carrot juice is both nutritious and delicious, but needs to be drunk with a little restraint. In 1974 Basil Brown from the United

Kingdom drank 10 gallons of the stuff in just 10 days. What that means is that he ingested 10,000 times the recommended amount of vitamin A. Unfortunately for him, this amount caused severe liver damage and his premature death. I guess at times there can be too much of a good thing.

Water is a benign substance, a fluid necessary for life, but it can be fatal if taken in too large of a quantity. In August 2014 a Georgia teen died after drinking two gallons of water and two more gallons of Gatorade during football practice. He suffered massive swelling around his brain from over-hydration. It is rare, but there are cases of death from over-hydration every year or so. Drinking too much water can throw off the body's balance of electrolytes, which can cause brain swelling and lead to seizures, coma, and even death.

Unlike the over consumption of water and carrot juice, which at times has been fatal, in the use and even abuse of cannabis over thousands of years of recorded history there has never been a single fatality! Nearly all medicines and substances have toxic and potentially lethal effects, "but marijuana is not such a substance. There is no record in the extensive medical literature describing proven, documented cannabis-induced fatality."[19]

In 1972, University of Virginia professor Richard Bonnie was the associate director of a commission appointed by President Richard Nixon to study marijuana. The commission said marijuana should be decriminalized and regulated. President Nixon rejected that idea.

Not only has hemp been proven safe for use, it also has proven medical uses. You can find the documentary film online called *Run from the Cure*. It is about Rick Simpson, a Canadian man who rediscovered cannabis as a cure for cancer and many other diseases. He cured his own skin cancer, basal cell carcinoma, using 100 percent natural THC containing hemp oil. The film contains a number of documented testimonies, as well as information about the plant.

The prophet Isaiah wrote, "*Woe unto them that call evil good, and good evil; that put darkness for light, and light for darkness; that put bitter for sweet, and sweet for bitter*" (Isaiah 5:20). The Apostle Paul told the Romans to "*abhor that which is evil; cleave to that which is good*" (Romans 12:9). Amos, speaking to Israel, said, "*Seek good, and not evil, that ye may live: and so the LORD, the God of hosts, shall be with you, as ye have spoken*" (Amos 5:14).

As you read the following stories, try to determine who is on the side of malice and evil and who is on the side of righteousness, virtue, decency, justice, and the LORD. In these stories, who was seeking good and who was cleaving to evil, and who was calling good evil and doing evil? Consider the Apostle Paul's instruction and wise counsel: "*Prove all things; hold fast that which is good*" (I Thessalonians 5:21).

Angela Brown, a mother of two, faced two years in prison and $6,000 in fines for saving her son's life (the charges were eventually dropped and she was not tried for any crime). What was her crime, her hideous act of evil that required retribution and punishment? She treated her son's debilitating condition with hemp oil, which actually worked. After just a few drops, Trey's pain melted away. But when the authorities found out, the sheriff's department confiscated the oil and arrested her, and the county prosecutors charged Angela with child endangerment and required that the child be put under state protection.

She used cannabis oil to treat her son's horrifying seizures, which were the result of him being hit in the head with a baseball. This had caused a traumatic brain injury and gave him great pain, daily migraines, and muscle spasms, culminating in depression. To ease his pain, Trey's parents tried 18 different medications over three years, but nothing really helped. Angela believes the uncontrollable outbursts and suicidal trendies he experienced were side effects of the drugs she was giving her son. Through research, she found out about hemp oil, and though it was illegal to use where she lived, she was determined to help end her son's suffering. Shortly after she started Trey on the cannabis oil treatments, he was able to return to school. However, when someone heard he was using cannabis oil, they reported Angela Brown to the authorities and she was arrested.

Journalist Matt Agorist published a blog post titled "This Minnesota Mom is Facing 2 Years in Jail for Saving Her Son's Life" on November 24, 2014, on the web site Free Thought Project:

> Apparently there are still some people in society who possess a certain level of *criminal ignorance who* would turn in a mother to the police for trying to help her son. [. . .]
>
> Those who would throw a person in jail for possessing a plant are criminals. Those who would throw a mother in jail for trying to help her suffering son have a special kind of iniquitous repugnance far beyond that of a criminal . . .[20]

I wholeheartedly agree with his comments. Those who arrested Trey's mother and took away the child's medicine are guilty of child endangerment. They are the ones who should be arrested and charged with those crimes. For the Christians reading this, I hope you will say a hearty "AMEN!" to Agorist's comments. Sadly, I know some of you still have the wool pulled tightly over your eyes and are unable to see where the real evil lies.

This is not the whole story; there is a real twist of irony to this saga. Minnesota is now the twenty-second state to approve medical marijuana. As of July 2015, it is legal to use hemp oil for medical purposes in Minnesota (recreational use remains illegal).

This is genuine insanity. One day, a certain activity can land you in prison; the next day, that same "illegal" act is made legal. Even though medical marijuana became legal and her son's condition was covered under its provision, at the time, the district attorney refused to drop the charges against Angela. This was an arbitrary and capricious, a reprehensible and perverse use of the law.

In a similar case, the state of Kansas charged Shona Banda with five criminal counts,[21] took her eleven-year-old son into state custody, and she faced 30-plus years in prison.

The medical community had nothing more to offer Shona Branda as she lay dying. She was suffering from Crohn's disease, an autoimmune inflammatory disease of the digestive tract that can cause severe gastrointestinal distress. Its symptoms include persistent diarrhea, bleeding, and fatigue. Since Crohn's disease is considered incurable, those who have been stricken with it face a life-long battle. It is usually not considered a fatal condition, but it caused so many problems for Shona that she needed numerous surgeries and was still dying. The last doctor she saw for another medical condition related to Crohn's told her that her body was rotting from the inside out and she was just minutes or days away from death.

Her husband had heard that smoking marijuana could help relieve some of her symptoms and urged her to try it. She had never smoked it before, so she reluctantly gave it a try. Unlike all the numerous drugs prescribed for her over the years for her condition, it gave her almost immediate relief, but did not cure her. Through her research, she found out about hemp oil. One day as she was treating her illness she accidently discovered how to produce the oil and began to make her own. Within a few days of taking it, the oil began to work its magic and within a few months, it restored her to full health. She even wrote an inspiring book, titled *Live Free or Die*, about how medical marijuana relieved her debilitating illness. So why is she facing over 30 years in prison? She was making hemp oil from cannabis, and the plant and oil are illegal to possess in Kansas.

Here is another recent story by reporter Kim Bellware of the *Huffington Post*:

> In 2015, the rate of absolutely zero deaths from a marijuana overdose remained steady from the year before, according to figures released in December by the Centers for Disease Control. [...]
>
> A total of 17,465 people died from overdosing on illicit drugs like heroin and cocaine in 2014, while 25,760 people died from overdosing on prescription drugs, including painkillers and tranquilizers like Valium, according to CDC figures. [...]

> The more than 30,700 Americans who died from alcohol-induced causes last year doesn't include alcohol-related deaths like drunk driving or accidents; if it did, the death toll would be more than two and a half times higher. [...]
>
> Though marijuana has yet to lead to a fatal overdose in the U.S.... but taking too much will likely lead to, if anything, a really bad trip.
>
> Despite the changing tide in American attitudes toward marijuana for both therapeutic and recreational uses, legalization is still vigorously opposed by groups like the pharmaceutical lobby (who stand to lose big if patients turn to medical marijuana for treatment) and police unions (who stand to lose federal funding for the war on drugs).[22]

The last paragraph in this story gives the reason some groups oppose legalization, certainly not the plant's safety, but their appetite for money, greed! They put money ahead of people's welfare.

On the Healthy Hemp Oil website is another true story, "How Children Brought Light to CBD: The Story of Charlotte," about a little girl named Charlotte Figi, who probably would not be alive today without hemp oil.[23] Many doctors and researchers will try just about any kind of medical experimentation on their patients, but God forbid they should use a medicine that has never killed anyone!

Now that you know a little about marijuana, ruminate on this for a while: Legal prescription drugs kill thousands of people a year, have numerous side-effects, and in most cases neither cure nor helps, but are still considered *good*. In most places, it is still illegal to grow hemp for making your own medicine. Marijuana has almost no known side-effects; it is not addicting and is one of the least toxic substances known to man. It has never killed anyone and has been used by millions of people for thousands of years to treat illness. It offers relief from disease, and the oil and other products that are made from this plant can help eliminate symptoms and, in many cases, cure disease. However, despite these facts, it is considered *bad* and thus is illegal. "*Woe unto them that call evil good, and good evil*" (Isaiah 5:20).

Before we continue, if you are still not sure where I stand concerning this plant, let me clearly spell it out for you: A few years ago I was completely unaware of the benefits and value of hemp. Because I was totally ignorant, I was biased, prejudiced, and intolerant of the use of anything that was produced from hemp. Because of the thorough job the media did in giving hemp an evil reputation, I believed anything associated with it was tainted, corrupt, and defiled. When I learned that clothing, food, and medicine could be made from hemp, an idea that was foreign to me, I had great reservations and qualms about using any of those products.

If one so chooses, education can be a cure for prejudice and bias. My change of heart came about because of my exposure to the plethora of surprising and abundant information from numerous, and at times unexpected sources about the benefits of hemp. While I do not promote the misuse of this plant, I am now an advocate and support the unrestrained growing and production of this plant and use of its many products. I also believe that *all* laws and regulations on the books that prohibit or limit the production and growing of hemp, restrain the sale of hemp products, curtail a country's importation or exportation of it, and make the use of hemp medicines illegal are evil, malevolent, criminal, and greatly harmful to all of mankind.

So why is this plant so despised and reviled by so many people? A unified campaign by a certain segment of society, a powerful and elite group of businessmen who oppose full legalization of hemp, has fuelled the negativity. With the help of complicit legislators, news media, and even the United Nations, they have been able to orchestrate and wage a successful worldwide campaign against hemp. For them, hemp is an evil competitor and any means available to hinder, obstruct, and impede its production and use is in their best interest.

In 2014 President Obama signed a bill (it was an amendment on the bill) to make the cultivation of hemp legal for commercial and research purposes, but only allows State Agriculture Departments, colleges and universities to grow non-drug oilseed and fiber varieties of cannabis for academic or agricultural research purposes. However, this amendment only applies to states where growing industrial hemp is already legal under state law (about half the states). This is a very small step in the right direction:

> despite state authorization to grow hemp, farmers in those states still risk raids by federal agents, prison time, and property and civil asset forfeiture if they plant the crop, due to the failure of federal policy to distinguish non-drug oilseed and fiber varieties of *Cannabis* (i.e., industrial hemp) from psychoactive drug varieties (i.e., "marihuana").[24]

What the Apostle Paul wrote to his young friend, convert, and minister is perfectly applicable to the businessmen in our discussion: "*But they that will be rich fall into temptation and a snare, and into many foolish and hurtful lusts, which drown men in destruction and perdition. For the love of money is the root of all evil...*" (I Timothy 6:9–10). Doug Yurchey put it this way:

> Marijuana does NOT pose a threat to the general public. Marijuana is very much a danger to the oil companies, alcohol, tobacco industries and a large number of chemical corporations. Various big businesses, with plenty of dollars and influence, have suppressed the truth from the people.
>
> In 1937, DuPont patented the processes to make plastics from oil and coal. DuPont's Annual Report urged stockholders to invest in its new petro-chemical division. Synthetics such as plastics, cellophane, celluloid, methanol, nylon, rayon, dacron, etc., could now be made from oil. *Natural hemp industrialization would have ruined over 80% of DuPont's business...* the real reason is that Big Business wants us to use petrochemicals and fossil fuels. They have little interest

in NATURAL solutions. THEY have suppressed the truth concerning cannabis and purposely created the 'menace of marijuana.' Why?

So they remain high profiteers while destroying our environment in the process...[25]

Yurchey summarized so well what the conspiracy has managed to accomplish during the last 80 years when he states: "They have created a monster out of the very thing that is a precious gift to the world." [26]

On August 18, 2015, Ethan A. Huff, a staff writer for the NaturalNews website, wrote an article titled "Why Big Pharma Hates Legalized Marijuana; Painkillers, Chemotherapy and Psych Drugs Could Be Made Obsolete":

When certain politicians and so-called health experts bellyache over the legalization of cannabis, it's not because they're worried about the children. It's because cannabis is arguably the safest and most widely effective natural medicinal herb in existence, and its widespread acceptance would immediately render obsolete the gamut of pharmaceuticals that make the establishment drug barons (and their bought-and-paid-for lackeys in Congress) filthy rich.

No longer would the general population need to rely on side effect-laden painkiller drugs, psych meds and other deadly pharmaceuticals for relief from their chronic ailments—cannabis would replace all this and more, providing true healing to the masses for pennies on the dollar. Because cannabis is a *plant* that anyone can grow, it also threatens the centralized power structures that control modern medicine, not to mention the for-profit prison industry that banks on incarcerating non-violent drug offenders.[27]

You know what they say about oats: If you want good-quality oats, they come at a premium price. However, if you can be satisfied with oats that have already gone through the horse, that comes a little cheaper. It seems many people today just accept the cheaper product, which in this case is what they hear on newscasts and read in newspapers. They never bother to "look behind the curtain," so they will never discover the deception that has been perpetrated on them. If you want good-quality, non-biased, factual information, you will not get it from the major news media sources, for after the news has gone through the "horse" it will come out filtered and pre-digested, an inferior product, just what they want you to feed on.

Jack Herer (June 18, 1939–April 15, 2010) was sometimes called the "Emperor of Hemp." He was an American cannabis activist and author of *The Emperor Wears No Clothes*. His book has been used in efforts to decriminalize and legalize cannabis and to expand the use of hemp for industrial use. The title of the book alludes to Hans Christian Andersen's classic fairy tale *The Emperor's New Clothes*. Herer uses Andersen's story as an allegory for the current prohibition of hemp.

After I read his book, I thought, "if only half of what he has written in this book is true, we have been fed a bunch of horse manure that has helped to impoverish the nations." The back cover summary says the following:

> If all fossil fuels and their derivatives, as well as trees for paper and construction were banned in order to save the planet, reverse the Greenhouse Effect and stop deforestation; then there is only one known annually renewable natural resource that is capable of providing the overall majority of the world's paper and textiles; meet all of the world's transportation, industrial and home energy needs, while simultaneously reducing pollution, rebuilding the soil, and cleaning the atmosphere all at the same time... and that substance is—the same one that did it all before—Cannabis Hemp... Marijuana![28]

Is it true what Herer and other researchers have written about the benefits and potential possibilities of the hemp plant? What have the critics said? Dr. Hayo M.G. van der Werf is an European author and expert on hemp. His doctoral thesis was titled "Crop Physiology of Fibre Hemp" (1994). He had this to say about Herer's book: "Although most of the information contained in the book is valid, some of its claims are clearly incorrect."[29] What an admission from an expert! Most of the information is valid! However, he criticized the book for containing, in his opinion, some inaccuracies. But what claims and information did he find to be incorrect? He mentioned nothing about the historical facts, data, or history of hemp or its numerous uses and medical befits. He only criticized Herer for making what he asserted to be unrealistic claims regarding the potential of hemp. He said some of the claims concerning the cellulose dry weight of the plant were incorrect and that under the most favorable growing conditions, there were other crops, such as maize, sugar beet, or potato, that could produce similar dry matter yields. Nevertheless, assuming Dr. Werf is correct, his criticism does not put much of a damper on the book as a whole, nor its main assertions about hemp's huge benefits and potential.

After reading all the information given concerning this plant, some may still be against its legalization, for they believe marijuana is a "gateway drug." On the *Time* web site is an October 29, 2010 article by Maia Szalavitz, "Marijuana as a Gateway Drug: The Myth That Will Not Die," which reads in part:

> Scientists long ago abandoned the idea that marijuana causes users to try other drugs: as far back as 1999, in a report commissioned by Congress to look at the possible dangers of medical marijuana, the Institute of Medicine of the National Academy of Sciences wrote:
>
> "Patterns in progression of drug use from adolescence to adulthood are strikingly regular. Because it is the most widely used illicit drug, marijuana is predictably the first illicit drug most people encounter. Not surprisingly, most users of other

illicit drugs have used marijuana first. In fact, most drug users begin with alcohol and nicotine before marijuana—usually before they are of legal age."

In the sense that marijuana use typically precedes rather than follows initiation of other illicit drug use, it is indeed a "gateway" drug. But because underage smoking and alcohol use typically precede marijuana use, marijuana is not the most common, and is rarely the first, "gateway" to illicit drug use. There is no conclusive evidence that the drug effects of marijuana are causally linked to the subsequent abuse of other illicit drugs.[30]

You can find many such stories on the Internet, both for and against. But as pointed out before, there is always a hidden agenda behind the prohibition of marijuana and keeping its use illegal. Follow the money trail: there is lots of dough to be made by hindering the competition and in pursuing and incarcerating those "law" breakers.

PSEUDOSCIENCE

I will conclude this chapter with our next topic. Perhaps you have heard the term *pseudoscience* but are not sure what it is. The term has been in use since the late eighteenth century in reference to alchemy. The etymology of the word is derived from the Greek root *pseudo*, meaning false, and the English word science. Thus, pseudoscience is *false science*. The 1989 edition of the *Oxford English Dictionary* defines it this way:

> A pretended or spurious science; a collection of related beliefs about the world mistakenly regarded as being based on scientific method or as having the status that scientific truths now have.

Pseudoscience has probably fooled most people at least once during their lifetime. Some people are misinformed their entire lives and never wise up or discover their ignorance. Ignorance about a partial topic or issue is not usually a life or death matter, and folks may go on with life without enduring the slightest consequences from their mistaken opinions. A belief in astrology, for instance, may cause someone to hold to strange ideas, such as relying upon the daily reading of their horoscope to help determine their course of action. Rarely, however, will those beliefs be the cause of earth-shattering events or actions.

Pseudoscience comes in many forms, constructs, and models. There may be systems and rituals that go along with each practice or its belief. Here is an abbreviated list from Wikipedia what many consider pseudoscientific concepts or ideas:

> Acupuncture, alchemy, ancient astronauts, applied kinesiology, astrology, biorhythms, cellular memory, cold fusion, craniometry, **creation science**, esoteric healing, extrasensory perception (ESP), graphology, homeopathy, intelligent design, iridology, metoposcopy, N-rays, naturopathy, paranormal plant perception, phrenology, polygraph, qi, New Age psychotherapies (e.g., rebirthing therapy), reflexology, reiki, Rolfing, therapeutic touch. (emphasis added)[31]

The intelligent design movement may be pseudoscience, but I am not going to discuss it here.

I may get some heat from fellow creationists for writing this, but I believe creation science is neither a scientific model nor a scientific discipline. While it is true that this model can be used to make predictions and it does not violate known scientific principles, each discipline listed above, including creation science, is probably outside the bounds of scientific investigation. Creation science deals with the origins of the universe, the creation of our planet and life, which are not in the purview of testability nor observation; rather, the main ingredient lies within the boundaries of faith and belief.

One criterion used to distinguish science from pseudoscience is the idea of falsifiability, which means something can be proved to be incorrect. What kind of tests could you perform that could prove or disprove the statement, "In the beginning God created the universe"? There is no scientific way to prove it true or false, for no tests can be devised to demonstrate it either way; it lies outside the reach of observation and so outside of science.

The five steps of the scientific method that scientists generally follow were discussed in Chapter 7. What is so interesting about this scheme and technique is that normally evolutionists agree with it, at least in principle, but they will not be consistent in applying it. Those who have taken a science class should remember one of the first things they learned: "*Science is based on observation.*" However, that fact is completely disregarded by anyone who claims that the *religion of evolutionism* is science. The study of history is not science because those who study it cannot observe, test, or experiment on the past.

In Wikipedia's abbreviated list of numerous types of pseudoscience, there is one glaring error: the "theory" of evolution was left off the list. The practice and pursuit of evolutionary theory is pseudoscience. Though presented as science, as we have seen, it fails to meet the norms of scientific investigation.

The religious belief of creation science was developed, for the most part, in response to the evangelistic efforts of the practitioners of the pseudoscience called evolutionism. Evolutionism is a jealous master that will admit no other ideology. Any perceived threat to its dogma is condemned and denounced as loudly and vehemently as possible. Through stealth and deceit, and with the help of lies found in all science textbooks, evolutionists have been able to banish creation from public school curricula.

Between these two beliefs, Creation and evolutionism, there exist fundamental conflicts, for they are rival religious systems. In choosing one, you reject the other. However, creation, unlike evolutionism, is in complete harmony with science and nature. This statement is confirmed by Dr. von Braun:

After years of probing the spectacular mysteries of the universe, I have been led to a firm belief in the existence of God. The grandeur of the cosmos serves only to confirm my belief in the certainty of a creator. I just cannot envision this whole Universe coming into being without something like divine will. The natural laws of the Universe are so precise that we have no difficulty building a spaceship to fly to the moon and can time the flight with the precision of a fraction of a second. These laws must have been set by somebody that "science" and "religion", properly understood, are not antagonistic pursuits... [On the contrary] they are "sister disciplines". Through the scientific method, one learns more about the "creation", whereas by virtue of the study of religion, one gains a greater insight into the "creator."[32]

Creation science and evolutionism are both pseudoscience. However, unlike creation science, evolution is a faith which holds contradictory views of nature and the universe, for its beliefs and tenants do not correspond with what we observe. While creation science is not science in the true sense of the word, it is a faith which is bound with science, holds to what is observed in nature, and is not at odds with the world God created. Evolution is a hoax—an evil faith held by the willfully ignorant. Ignorance may indeed draw the presumption that reindeer fly, but reason and true science come to a different conclusion.

NOTES ON CHAPTER 9

1. "Cardiff Giant," Wikipedia web site, https://en. wikipedia. org/ wiki/ Cardiff_Giant
2. "There's a Sucker Born Every Minute," Wikipedia web site, https://en.wikipedia.org/wiki/There%27s_a_sucker_born_every_minute para. 3.
3. Ibid. Another very interesting and detailed account of the Cardiff Giant hoax can be found in "P.T. Barnum Never Did Say 'There's a Sucker Born Every Minute,'" on HistoryBuff.com.
4. "Dihydrogen Monoxide Hoax," Wikipedia web site, https:// en. wikipedia. org/wiki/Dihydrogen_monoxide_hoax.
5. Ibid., para. 10.
6. Much of the information about this plant presented here was taken from the book *The Emperor Wears No Clothes* by Jack Herer. What is in this chapter is just a small fraction of the information contained in Herer's book. You can read it for free online at http://jackherer.com/emperor-3/.
7. "The Ford Made of Hemp," Hidden Mysteries web site, http://www.hidden-mysteries.org/conspiracy/facts/fordhemp.html, paras. 2–4.
8. Herer, n.d., Chapter 3: Textiles and Fabrics, para. 1.
9. Lance D. Johnson, "Dangerous Medicine: America's Opioid Epidemic Isn't the Only Prescription Drug-Based Crisis Killing Hundreds of Thousands," Natural News web site, November 24, 2017, https://www.naturalnews.com/2017-11-24-dangerous-medicine prescription-drugs-are-still drugs.html, para. 3.
10. For details on tobacco and alcohol deaths, see: CDC, "Alcohol and Public Health: Fact Sheets – Alcohol Use and Your Health," CDC web site, https://www.cdc.gov/alcohol/fact-sheets/alcohol-use.htm, January 3, 2018; CDC, "Smoking and Tobacco Use: Tobacco-Related Mortality," CDC web site, December 1, 2016, https://www.cdc.gov/ tobacco/ data_ statistics/fact_sheets/health_effects/tobacco_related_mortality/index.htm

    Prescription drugs: 100,000. Taken as directed, Melody Petersen, author of *Our DailyMeds*,

    https://www.alternet.org/story/ 147318/100 % 2C000 _americans_die_each_year_from_prescription_drugs%2C_while_pharma_companies_get_rich.

    For information on aspirin deaths: see "Death by Aspirin," *What the Doctors Don't Tell You*, 20(1), https://www.wddty. com/magazine /2009 / april/death-by-aspirin.html

    In the first of the new studies, researchers reckoned that the drug is killing 20,000 Americans-and sending another 100,000 to hospital-every year just because of GI problems. Astonishingly, each of these deaths and reactions is slipping under the radar, and is not being associated with aspirin at all. (para. 7.)

If the researchers' death and injury rates for the US is [*sic*] correct, this could mean that the worldwide aspirin and other NSAID death toll is 100,000 individuals every year. What's more, a further 500,000 of us will need hospital care to treat a serious reaction to the drug, such as the result of a GI reaction (Proceedings of the Annual Scientific Meeting of the American College of Gastroenterology, 15 October 2007). Currently, none of these deaths is even recorded as an aspirin fatality. (para. 10)

For information on cocaine and heroin deaths, see: "How Many People Die from Drugs, Each Year?" Promises web site, https://www. promises. com/ resources/ overdose/many-people-die-drugs-year/ [2015 statistics].

- Heroin ... 12,982 deaths [...]
- Synthetic opioids such as fentanyl and tramadol ... 9,580 deaths [...]
- Overdose deaths involving methadone ... 3,285 deaths [...]
- Cocaine-related overdose ... 6,784 deaths [...]
- Benzodiazepine-related overdose deaths ... 8,791 (para. 4).

For information on hemp deaths, see: "Deaths from Marijuana v. 17 FDA-Approved Drugs (Jan. 1, 1997 to June 30, 2005)," State of Oregon web site, n.d., http://www.oregon.gov/Pharmacy/Imports/Marijuana/Public/Deaths From-MarijuanaV17FDAdrugs.pdf.

11. "Deaths from Marijuana ..." The Substance Abuse and Mental Health Services Administration's (SAMHSA) 2003 report *Mortality Data from the Drug Abuse Warning Network* ... stated: "Marijuana is rarely the only drug involved in a drug abuse death. Thus ... the proportion of marijuana-induced cases labeled as 'One drug' (i.e., marijuana only) will be zero or nearly zero" (p. 6).
12. Joseph Cariz, "Is it True that Marijuana Hasn't Caused Any Deaths, While Prescription Drugs Have Caused 100,000?" Politifact web site, http://www.politifact.com/truth-o-meter/statements/2016/aug/16/gary-johnson/gary-johnson-claims-marijuana-cant-kill-and-prescr/, paras. 9,10, 12, 13, 16.
13. "New Billion-Dollar Crop," *Popular Mechanics*, February 1938, 238–9.
14. "The Most Profitable and Desirable Crop that Can Be Grown," *Mechanical Engineering*, February 27, 1937.
15. Herer, n.d., Chapter 2: Textiles & Fabrics section, para. 14.
16. Doug Yurchey, "The Marijuana Conspiracy: The Real Reason Why Hemp is Illegal," World Mysteries web site, n.d., http://old.world-mysteries.com/marijuana1.htm, para. 9.
17. "Hemp for Victory!" [film], U.S. Department of Agriculture, 1942, at 0:36.
18. Marijuana Safety—DEA Administrative Law Judge's Ruling. U.S. Department of Justice, Drug Enforcement Administration, "In the Matter of Marijuana Rescheduling Petition" (Docket #86-22), September 6, 1988, p. 56–57.
19. Ibid.
20. Matt Agorist, "This Minnesota Mom is Facing 2 Years in Jail for Saving Her Son's Life," Free Thought Project web site, November 24, 2014, http://.thefreethoughtproject.com/Minnesota-mom-facing-2-years-jail-saving-sons-life/, paras. 7, 9.

21. At the time of this writing, with the help of Attorney Jennifer Ani, Shona has regained custody of her son, but her criminal trial has yet to begin.
22. Kim Bellware, "Here's How Many People Have Fatally Overdosed on Marijuana," *The Huffington Post* web site, April 20, 2016, http://www.huffingtonpost.com/entry/marijuana-overdose-deaths_us 5716468ee4b 0060ccda452ad, paras. 2–3, 6, 8–9.
23. Vanessa Benoit, "How Children Brought Light to CBD: The Story of Charlotte," Healthy Hemp Oil web site, n.d., https:// healthyhempoil. com/cbd-for-kids/
24. "President Obama Signs Farm Bill with Amendment to Allow Industrial Hemp Research," Cision web site, February 7, 2014, https:// www. prnewswire.com/news-releases/president-obama-signs-farm-bill-with-amendment-to-allow-industrial-hemp-research-244342691.html
25. Yurchey, n.d., paras. 8, 16.
26. "The Marijuana Conspiracy," World Mysteries Blog, http://blog.world-mysteries.com/science/the-marijuana-conspiracy/ PS section, para. 11.
27. Ethan Huff, "Why Big Pharma Hates Legalized Marijuana: Painkillers, Chemotherapy and Psych Drugs Could Be Made Obsolete," Natural News web site, August 18, 2015, https://www. naturalnews. com/050828_marijuana_Big_Pharma_natural_remedies.html, paras. 1–2.
28. Herer, n.d.
29. Hayo van der Werf, "Hemp Facts and Hemp Fiction," Drug Library web site, n.d., http://druglibrary.net/olsen/HEMP/IHA/iha01213.html, para. 1.
30. Maria Szalavitz, "Marijuana as a Gateway Drug: The Myth that Will Not Die," Time, October 29, 2010, http://healthland.time. com/ 2010/10/29/marijuna-as-a-gateway-drug-the-myth-that-will-not-die/, paras. 4–5.
31. "Category: Pseudoscience" [directory page], Wikipedia web site, n.d., https:// en.wikipedia.org/wiki/Category:Pseudoscience.
32. Chester Delagneau, "Integration of Science and Religion by Dr. Wernher von Braun" [Personal blog], August 30, 2017, http:// chesterdelagneau. com/?p=36.

It is obvious that radiometric techniques may not be the absolute dating methods that they are claimed to be. Age estimates on a given geological stratum by different radiometric methods are often quite different (sometimes by hundreds of millions of years). There is no absolutely reliable long-term radiological "clock". The uncertainties inherent in radiometric dating are disturbing to geologists and evolutionists...'.*

CHAPTER 10

DATING ROCKS AND DEAD THINGS

And the earth was without form, and void; and darkness was upon the face of the deep. And the spirit of God moved upon the face of the waters. (Genesis 1:2)

A SCIENTIST WALKED INTO A large ornate office building. It had been built decades before by an eccentric millionaire. The cathedral ceiling in the atrium was exquisite; it was covered with murals that reminded the scientist of the Sistine Chapel frescos, scenes from Genesis painted by the hand of the master, Michelangelo.

In a uniquely themed, matching centerpiece, a burning candle of considerable size rested in a prominent spot. This flickering candle was beautifully crafted; it was a blend of inlays and veneers, a plethora of colors that were skillfully woven together, a masterpiece that gave off a mesmerizing, soft, translucent glow that enveloped the whole chamber.

---
*William D. Stansfield, Ph.D. (animal breeding) (Instructor of Biology, California Polytechnic State University) in *The Science of Evolution*, Macmillan, New York, 1977, p. 84.

He thought it a shame that one day its beauty would be totally devoured. While he wondered how long it would be before its demise, he became fixated on the question of how long had it been burning and how tall it was when it started to burn. He asked a number of concierges, but being newly hired, not one of them had any knowledge of when it had been lit. He was able to speak to the owner, but he told the inquisitive scientist that he had recently purchased the building and the candle was burning at the time he obtained ownership. Compounding the problem, he was not able to reach the previous owner and the former manager had moved to parts unknown. Furthermore, all the invoice(s), tax records and transactions dealing with the candles' origin were stored on microfiche, but had been lost in a fire more than a decade ago.

Being recently retired from heading up a research foundation, though still busy, he allowed himself a full week to satisfy his irrepressible curiosity. So, using the skills acquired while studying for his Ph.D. and work experience, he put them to use by running scientific tests. Using these tests, he hoped to determine the length of time the flame had been consuming the wax, thereby reducing the elevation of the candle and revealing its original height. First, he carefully measured the current height of the candle and its circumference. He found that its diameter was consistent throughout its entire length. He then took a sample of the wax and had it analyzed. It was determined to be beeswax of recent origin. He then compared its properties against other types of wax. He examined the wick, measured the temperature of the flame and how many lumens it put out, and took readings of the ambient air in the room.

Next, he charted the rate of time it took for the flame to consume the wax. At its present rate, he could determine the precise amount the candle would burn down per hour. Then, by extrapolating the current rate of wax consumption, he could precisely determine its remaining height at any given point in the future. Still, he had not found the answers to his quest. He continued to devise tests, take measurements, ponder, deliberate and muse over this problem. But after his week of analysis, investigation, and evaluation, he was no closer to discovering either the original height of the candle or the length of time it had been burning.

He could however, make a few conclusive determinations. First, it seemed obvious that the original height of the candle could not have been taller than the ceiling. Thus, using the ceiling as the apex of the candle when it started to burn, he could now determine the length of time it had been burning. However, how was he to know if the height of the ceiling was the actual starting point? For its original acme could have been at any point between its current status and the ceiling.

Using the scientific method along with mathematical equations and calculations, many times an unknown quantity can be determined, a problem solved, or a reasonable guess can be made. However, in this case, even with all the information that was now at his disposal and using his scientific expertise, unless he could track down the artist who created this magnum opus and the person who lit it, the original height of the candle and the length of time it had been burning are impossible to determine.

There is one further twist or irony to this story which applies to the study of the unknown—to past ages in general and fossils specially. There are things concerning the past that researchers will never fully understand, no matter the amount of time spent on research, investigation, or examination. Things evolutionists now assume and consider as fact, someday in the distant or near future, some may awaken, and discover they were totally and completely mistaken.

Even with all of the information now known by our unnamed curious science researcher, and using the height of the ceiling as the absolute zenith of the candle when it started to burn, even his educated guess could never come close, nor was it even possible to determine the original height of the candle when it commenced to burn. Unbeknownst to the present owner of the building, the current staff, and by our researcher, the artist, recently diseased, who created this masterpiece, was also an inventor and skilled mechanic.

Unobserved by the scientist, a considerable length of candle was hidden, for it protruded through a hole which was cunningly concealed in the floor. As the candle burned down, once a week, a precisely timed mechanical device slowly raised the candle the exact amount it had burned down [the scientists' timing was unfortunate, he began his study just after a new section of candle was added]. Thus, it never seemed to burn up. This process could continue for many decades, for ingeniously conceived and designed, replacement sections of candle would be as needed, melted to the bottom of the candle by a heating element, grafting it to the bottom of the remaining part, so that it fit seamlessly. Only when all of the replacement sections were used up could the candle truly burn down.

Our researcher had based his conclusions on a faulty starting point, assumptions which were wrong, and a total misunderstanding of how the process worked; the same as evolutionary "researchers" with their dating schemes (any radiometric dating must always fit within the bounds of evolutionary doctrine of long ages). Christians on the other hand are privy to some information evolutionists disregard and ignore. We do understand who created, when, and why all things came into being. Keep these things in mind as we consider this next section.

RADIOMETRICITY, THE ART OF DATING ROCKS

Before we discuss this subject, let us ponder a few things. Just how does a planet form? After the big bang, all its created material would have been flung in all directions; with every second that passed it would have become more dispersed, less dense, and scattered. In the beginning, when the big bang took place, there would have been no magnetic or gravitational fields or friction in space, so there would have been nothing to hinder, slow, or alter an objects course after its initial acceleration (for space is a near perfect vacuum), and thus, it would travel at great speed forever, in the same direction. When the singularity (the big bang) exploded, each piece of debris would move at right angles away from its source. Even two pieces of debris in close proximity to each other, because of the curvature of the surface of the exploding sphere, would immediately begin to increase the distance from each other. And the longer they traveled from their point of origin, the father apart they would become; thus, they could never bump into each other. So, how does a loosely packed cloud of dust and rocks traveling at mind-numbing speeds, each partial moving away from its point of origin, change course, bump into each other, and stop long enough to coalesce and fuse together to form stars, moons or planets? Certainly none of this was observed. Here is one of their ideas:

> The Earth is thought to have been formed about 4.6 billion years ago by collisions in the giant disc-shaped cloud of material that also formed the Sun. Gravity slowly gathered this gas and dust together into clumps that became asteroids and small early planets called planetesimals. These objects collided repeatedly and gradually got bigger, building up the planets in the Solar System, including the Earth. The details of how the Earth formed are still being worked out. Scientists study meteorites and the oldest rocks on Earth to understand what happened in these earliest times in the Solar System. They also observe other solar systems in our galaxy, the Milky Way.[1]

This is not a theory or a hypothesis; it is just a silly story, a wild guess built firmly on faith in evolutionism. Two objects traveling at high rates of speed do not clump together when they meet, but explode, creating smaller pieces, which then fly in all directions. Anyway, after much bumping and jostling of space junk, which formed the Earth, it was a bit on the warm side:

> In the very beginning of earth's history, this planet was a giant, red hot, roiling, boiling sea of molten rock—a magma ocean. The heat had been generated by the repeated high-speed collisions of much smaller bodies of space rocks that

continually clumped together as they collided to form this planet. As the collisions tapered off the earth began to cool, forming a thin crust on its surface. As the cooling continued, water vapor began to escape and condense in the earth's early atmosphere. Clouds formed and storms raged, raining more and more water down on the primitive earth, cooling the surface further until it was flooded with water, forming the seas.[2]

So, over the course of billions of years, at some point during these creative events and activities, radioactivity was given birth. What caused some elements to be radioactive and some to be stable? At just what point did unstable atoms start releasing energy? How long does it take for an element to become radioactive? Was it an instantaneous happening, like flipping a light switch? Or was it a long, drawn-out affair, accomplished over the course of millions of years? Was each element 100 percent pure when it was created? How big was each element in the beginning when it became radioactive? If the element was large, say the size of a small island, isn't it possible some parts might not have been radioactive? Kind of like that chicken, the master of the coals (your dad) barbecued during one summer picnic, it was fully cooked on the outside, but still raw in the middle.

Now, there may be a scientific explanation for one or two of my questions dealing with radioactivity. But this whole story rests upon something as solid as quicksand, just as deceptive and disingenuous. And upon this yarn evolutionists insist they can build a reliable mathematical system, formula, or equation that will date rocks with great accuracy. So many unknowns and uncertainties, but they assure us they have cracked the code! Now, no matter how precise equations may be, if they are anchored upon a fairy tale, its analogous to working out the speed in which Santa's reindeer must fly to complete their rounds on Christmas Eve, the end result is worthless. But there is a reliable eyewitness that was there, for He was the Creator, and He tells us that in the beginning the Earth was covered with water, and not a "red hot, roiling, boiling sea of molten rock." Remember this information as we dig further into this issue.

Mistakenly, many assume that carbon-14 is one of the methods used to date fossils. Most fossils form when the original organic matter in an organism dissolves slowly and is replaced, bit by bit, by chemicals or minerals that gradually seep into the bone and form a rock-like copy of the original. Normally, only the hard parts of an animal, such as the skull, teeth, or bones become fossilized.

Radiocarbon dating was pioneered by Willard Libby and his colleagues in 1949 as a method to date geological, archaeological, and hydro-geological samples. Its use is strictly limited to materials of an organic nature, such as cloth, bone, wood and plant fibers. Typically, fossils such as dinosaur bones do not contain any organic matter; radiocarbon could therefore not be used to date most fossils. It decays with a half-life of about 5,730 years, so its range is limited to about 50,000 years (much too short for the millions of years required for evolution), the amount of carbon left after that amount of time would be too small to analyze.

Radiometricity, or *radiometric dating*, is the science of determining the age of fossils, sediments, and rocks using radioactive isotopes in the rocks themselves. Using this method, scientists are confident they can generate absolute dates. "By measuring the amount of radioactive decay of a radioactive isotope with a known half-life, geologists can establish the absolute age of the parent material."[3]

In a question and answer format this information is stated:

> Do you believe radiometric dating is an accurate way to date the earth?
>
> Yes! Absolutely. It is an accurate way to date specific geologic events…There are many radiometric clocks and when applied to appropriate materials, the dating can be very accurate. […]

When an unstable element such as a Uranium (U) isotope decays, it turns into an isotope of the element Lead (Pb). The original unstable isotope, Uranium, is called the *parent* and the product of decay, the Lead, is called the *daughter*. From careful physics and chemistry experiments, we know that parent elements turn into daughters at a very consistent, predictable rate.

A geologist can pick up a rock from a mountainside somewhere, bring it back to the lab, and separate out the individual minerals that compose the rock. He or she can then look at a single mineral and, using an instrument called a mass spectrometer, measure the amount of parent and the amount of daughter in that mineral. The ratio of the parent to daughter then can be used to back-calculate the age of that rock.

The reason we know that radiometric dating works so well is because we can use several different isotope systems...on the same rock, and they all come up with the same age. This gives geologists great confidence that the method correctly determines when that rock formed. [...]

Scientists can measure the ratio of the parent isotopes compared to the converted isotopes. Because the rate of conversion of isotopes is known (how long it takes for a particular isotope to convert/decay), we can use the ratio to determine how old the object is that contains them.[4]

This radiometric method of dating materials we are discussing sounds so precise and accurate, so scientific, how could I or anyone doubt its accuracy? While the science that determines the rate of radioactive decay seems solid, when dating any material there is always a large unknown, a piece of the puzzle that will forever remain uncertain, indecipherable, and undiscovered. It is this unknowable and ambiguous information which will forever keep this method of dating in the realm of religious faith and belief, not science (please do not confuse the dating aspect we are discussing with other uses of isotopes).

If you recall, we are dealing here with points listed in Chapter one numbered 1, 2, and 3 from Dr. Hovind's lectures: "that adherents must 'believe' in since they have NEVER been observed or demonstrated" (emphasis his)—and, I would add, can never be tested.

When attempting to date something using a radioactive element, a scientist enters the picture after the countdown has already started. He or she has no way of knowing the original ratio of parent to daughter, how many parent and daughter isotopes there were when the process began. Without this critical unknown piece of information it is literally impossible to date anything from the past using radioactive elements.

This is similar to the situation of scientist who wanted to find out how tall and how long the candle was when it started to burn. All the analyzing in the world was not able to give him this crucial information. But all is not lost for the evolutionist, for this is where faith and belief, not science, comes into play. They make an assumption and take an enormous wild ridiculous guess founded entirely upon their religion:

> ... I should mention that the decay constants (basically a value that indicates how fast a certain radioactive isotope will decay) for some of these isotope systems were calculated *by assuming* that the age of the earth is 4.56 billion years, meaning that we will also calculate an age of 4.56 billion years if we use that decay constant... [...]
>
> *We assume* that the Earth is probably as old as the asteroids, because *we believe* the solar system to have formed from a collapsing nebula... The building blocks that the Earth is made of, the asteroids are 4.5 billion years old, and *we presume* that the Earth formed fairly quickly thereafter... [...]
>
> Based on astronomical models of how stars work, *we also believe* the Sun to be about 4.6 billion years old, slightly older than the rocky bodies in the solar system... (emphasis added)[5]

As we read through this story we find these words: "*by assuming that the age of the earth is 4.56 billion years . . . We assume . . . we believe . . . we presume . . . we also believe . . .*" This is the bedrock of their science, the facts they rely upon to build the case for radiometric dating. Now their unknown, the necessary piece of the puzzle needed to date materials becomes known, it is their *belief*, their *faith* in evolutionism, and *not science*, that magically materializes the needed starting point, the decay constant.

One evolutionist, after reading this, maintained that I was misreading the text, that the assumption only refers to current techniques, which cannot narrow down the age to how many million; thus it might be 4.53 billion or 4.59 billion. No, I did not misread the text; they clearly made a faith-based evolutionary assumption by assuming that the age of the earth is 4.56 billion years.

The truth is, evolutionary assumptions drive their age assignments. Their claim that objectivity and science confirm that rocks are millions of years old crumbles under scrutiny. As adjunct and assistant professor Brian Thomas[6] has stated:

> Secular scientists often assign ages to dinosaur-containing rocks by using radioisotope age assignments given to ancient lava flows found above or beneath them. But in order to convert measurements of radioactive versus stable atoms into ages, *one must assume a constant history for these crystalline rocks.*
>
> How accurate is such an assumption? First, consider radioisotope ages assigned to rocks of known age. (emphasis added)

Thomas then gives us an example using a chart:

Location	Known Age	Isotope Age
Mt. Erebus	17 yrs	1.6 my
Mt. Etna basalt	29 yrs	35 my
Mt. Etna basalt	37 yrs	0.7 my

Mt. Stromboli	38 yrs	2.4 my
Kilauea Iki	40 yrs	8.5 my
Mt. Lassen plagioclase	85 yrs	11 my
Kilauea basalt	<200 yrs	21 my
Hualalai basalt	200 yrs	0.6 my
Sunset Crater basalt	950 yrs	27 my
Mt. Etna basalt	2,100 yrs	25 my

my= millions of years

Thomas continues, pointing out another recent example:

> ...in 1996 Dr. Steve Austin described radioisotope results from crystalline rock collected at Mount St. Helens. Witnesses saw the lava harden to rock in 1986, six years after the famous 1980 eruption. The secular lab reported rock "ages" of about 0.35, 0.34, and 2.8 million years. Two million years for a ten-year-old rock? Something's very wrong here.

He sums it up this way:

> It turns out that the flaws in radioisotope dating do not arise from inaccurate isotope counts but from overconfidence in the accuracy of the assumption needed to convert those counts into "ages."

In July of 2015, two news stories appeared: "Rare Fossils of 400-Million-Year-Old Sea Creatures Uncovered" and "Ancient Comb Jellies Had Skeletons, but They Still Lost the Arms Race."[7] These stories concerned ancient creatures that supposedly lived over 400 million years ago in the Cambrian and Ordovician periods. Though there was a wealth of information contained within those two articles, there was nothing of substance, real science, or knowledge that could help us understand the past.

How do paleobiologists know the approximate time frame fossilized horseshoe crabs, jellies, and arthropods, such as anomalocaridids, cheloniellids and marrellomorphs mentioned in the stories, lived? Unless evolutionists have found an ancient Resister of Deeds archive that contains certified copies of these organisms' birth certificates, the ages advocated but stated as fact, are nothing more than an *uneducated* guess. It is like throwing darts at large numbers on a moving dartboard while blindfolded, which board is located two hundred yards away.

As no one observed these creatures when they were alive, what kind of scientific tests did the paleobiologists perform on these fossils to confirm their approximate age? Scientific tests do not establish the age of a fossil. Was an autopsy performed on these fossils to determine their age or potential relationship with other similar creatures? Even if an autopsy could be carried out, it could not be used to determine a fossil's age or any potential relationship between it and other similar creatures. Did they use any type of radiometric dating technique to substantiate their statements, the fossils age, or any other information presented in these news stories? In the early days of the study of geology, neither radiometric dating nor any other kind of scientific tests were made to determine the age of a fossil or its attendant ages. However, today, at times, radiometric dating techniques are used to try and determine the absolute ages of rocks and fossils. However, as has been pointed out, any date found is always subservient to the dates already assigned to the geological column which had been established years before the discovery of radioactivity or other dating methods.

In his article discussing the dating of fossils and the use of circular reasoning, R. K. Sepetjian quotes J. E. O'Rourke:

> A fossil cannot be dated by itself; it can only be dated by knowing where, in the geologic column, it was found. A layer of rock cannot be dated without knowing what fossils are found in it. This is why we cannot accurately date anything using the geologic column. "The rocks do date the fossils, but

the fossils date the rocks more accurately. Stratigraphy cannot avoid this kind of reasoning if it insists on using only temporal concepts, because circularity is inherent in the derivation of time scales."[8]

The system or method used to determine the age of the bulk of fossils found is the geological column.[9] Some of the history of this graph can be found on the Actionbioscience web site. Posted in January 2001, Michael Benton's article "Accuracy of Fossils and Dating Methods" said:

> Fossil sequences were recognized and established in their broad outlines long before Charles Darwin had even thought of evolution. Early geologists, in the 1700s and 1800s, noticed how fossils seemed to occur in sequences: certain assemblages of fossils were always found below other assemblages. The first work was done in England and France... geologists began to build up the stratigraphic column, the familiar listing of divisions of geological time—Jurassic, Cretaceous, Tertiary, and so on. Each time unit was characterized by particular fossils... From the 1830s onwards, geologists noted how fossils became more complex through time. The oldest rocks contained no fossils, then came simple sea creatures, then more complex ones like fishes, then came life on land, then reptiles, then mammals, and finally humans. Clearly, there was some kind of "progress" going on. All became clear, of course, in 1859 when Charles Darwin published his "On the origin of species". The "progress" shown by the fossils was a documentation of the grand pattern of evolution through long spans of time.[10]

Andrew MacRae, writing on The TalkOrigins Archive website, quoted: "By the end of the 1830s, most of the presently-used geologic periods had been established based on their fossil content and their observed relative position in the stratigraphy..."[11] Of course, between the 1840s and 1960s new classifications continued to be redefined and new eras named.

What scientific method, process, or technique was used to create the geologic column in the 1700s and 1800s? None. As Michael Benton's article clearly points out, the column simply came from a mistaken belief that the earth was ancient, that life formed spontaneously and then took a long time to evolve, from simple, into the numerous complex living organisms we see today. The whole evolutionary dating system is perched, floating, and suspended upon the geologic column, which itself is composed of nothing of substance or science. The column rests squarely upon a mistaken belief that simple organisms once existed, a myth produced by a flawed analysis of the strata that is bolstered by pure imagination, faith, and belief in the "progress" of life, evolutionism.

In an attempt to refute my contention that the scientific method and technique was not used to create the geologic column one evolutionist discussed Alexander von Humboldt (1769–1859), a man who was meticulous in his observational techniques, systematic measurements, and his use of every scientific technique he had access to. While this might be true, his research methods may have been almost flawless; they were based upon a much less than perfect understanding. As I got to know him, this conclusion proved accurate. Joseph McCabe writes this about him:

> One of the most encyclopedic scientists of the time, Humboldt was a Pantheist like his friend Goethe, and a contemptuous anti-clerical like his friend F. Arago ... His letters use[d] very strong language about the Churches to the end of his life. He calls Luther "that diabolical reformer."[12]

If you start your research with evolutionary ideas or flawed viewpoints, no matter how meticulous your research is, a faulty, unscientific, and unsound result will occur when you present your thesis.

Decades before Marie Curie (November 7, 1867–July 4, 1934) conducted her pioneering research on radioactivity or won the Nobel

Prizes in Physics and Chemistry, the geologic column was already in service and being used to assign fossils their respective ages, place, and order in the evolution of life.

> One can hardly accuse these pioneers of evolutionary prejudice. Nearly a half-century would pass before Darwin's book, *The Origin of Species*, was published! *By then, the relative ages (order) of the geologic column had already been worked out in some detail.* Radiometric dating would later confirm the relative ages of the strata and tie them to absolute dates. (emphasis added)[13]

Ernest Rutherford, a New Zealand-born British physicist and the "father of nuclear physics," studied radiometric dating during the early part of the twentieth century. But in 1907, Bertram Boltwood was the first person to publish material on the use of radiometric dating. A few years later, in 1911, Arthur Holmes seemed to be the first to develop a dating method based on radioactive decay, and sometime around the 1920s, the first radiometric dates were generated. Since then, scientists have continued to repeat the tests, upgrade and recalibrate their equipment, and have found better and more sophisticated techniques to improve and enhance their craft.

Among the types of radiometric methods used to date rocks and other materials are rubidium/strontium, thorium/lead, potassium/lead, and uranium/lead. Using these and other methods, the geologic time scale is updated every few years, and the major timelines and older dates may change a few million years accordingly. But no matter how precise, scientific, or accurate their tests are or may seem to be, they are simply cosmetic, for all their findings and conclusions are predetermined by the boundaries established in Darwin's day. All ages assigned, findings, and results must fall within the limits, constraints, and parameters ordained by the geologic column, which rests firmly upon a conviction that "once upon a time slime plus time produced Einstein."

LIVING FOSSILS

Tom DeRosa is a man well-qualified to level criticism and condemnation against those in the evolutionary community.[14] He has said: "Evolution is only made to appear credible in the fossil record by tremendous amounts of bias, manipulation and outright censorship employed by secular paleontology." The results from evolutionary dating methods are full of enigmas and problems. Hundreds—probably thousands—of creatures, plants, and animals previously thought to be extinct for millions of years have been found alive. Evolutionists call these finds *living fossils*. One of the most recognized and well-known living fossil is the coelacanth. Fossils of this fish that supposedly date back over 350 million years have been found. They were once known only by their fossils and thought to have gone extinct the same time as the dinosaurs, approximately 65 million years ago.

In December of 1938, Captain Gossen, a fisherman, caught a live coelacanth near East London, South Africa. A second specimen was captured in 1952 near the coast of the French-controlled Comoros Islands, which are located off the eastern coast of Africa near Madagascar. Since its rediscovery, numerous expeditions have set out to expand human knowledge of this lobe-finned fish nicknamed "old fourlegs." Since that time, many live coelacanths have been sighted and photographed off the coasts of Mozambique, Tanzania, Kenya, north of Papua New Guinea, and in the Western Pacific Ocean.

A more recent find is the Wollemi Pine (Binomial name: *Wollemia nobilis*), discovered in 1994 in Wollemi National Park in New South Wales, a remote temperate rainforest wilderness 150 km northwest of Sydney, Australia. The oldest known fossils of this tree have been dated to 200 million years ago. This towering tree—40 m high, 3 m in girth—was not trying to elude capture or discovery by man, but was taking pleasure in solitude, simply minding its own business and enjoying the company of its family, which at the time of discovery was made up of 23 adults and 16 juveniles. It also reveled in its daily routine of showing off its knobby, dark-brown bark, which resembles Coco-Puffs, and basking in the sunlight, sucking up water and nutrients from the soil, as it had been doing for hundreds of years.

The tree was rediscovered by David Noble, a project officer with the Australian National Parks and Wildlife Service. He was canyoneering a deep, rugged gorge in the Blue Mountains when the big tree caught his eye. He brought a branch back to show Wyn Jones, a senior naturalist with the NPWS. As they say, "the rest is history." Professor Carrick Chambers, director of the Royal Botanic Gardens, said, "It is a really major find. The discovery is the equivalent of finding a small dinosaur still alive on Earth." The scientific director at the gardens, Barbara Briggs, said, "On the world scene it's one of the most outstanding discoveries of the century."

Other rediscoveries of organisms back from the dead include a lungfish (fossil record: 385 million years old), discovered in 1870 by Burnett and Mary Rivers, and the Dawn Redwood (fossil record: 70 million years old), found in 1945 in the Sichuan province of China. The list of so-called living fossils is long and varied. It includes such creatures as the nautilus (a marine mollusk), the Ginkgo Biloba tree (a tree native to China used to make herbal medications for treatment of a number of disorders), the platypus, and the rarely seen and elusive goblin shark. It also includes the horseshoe crab, a quintessential creature that looks like something out of an artistic rendering of the Cambrian Explosion.

On Tom DeRosa's *Creation Studies Institute* webpage he relates some information concerning living fossils from the superb DVD by Dr. Carl Werner:

> Did you know that flamingos, sandpipers, penguins, cormorants, parrots, owls and many other creatures living today, including numerous types of mammals, reptiles, amphibians and arthropods are found in supposedly 65-plus-million-year-old rock layers, when dinosaurs and other "pre-

historic" beasts once roamed the earth? So, why don't we ever see them displayed in our museums or depicted together in books and textbooks?

Even though many fossils of these kinds of extent [sic] (or living) fauna have been found in allegedly "ancient" rock layers, this information is purposely censored from public view because it doesn't fit the evolutionist's millions-of-years timeline. The idea that animals we see today on a visit to the zoo occupied a spot in time with dinosaurs and earlier is considered "preposterous" by the chief evolutionary biologists and paleontologists at our universities and schools of higher education—even in the face of clear proof.[15]

Although some evolutionists might not deny that some modern animals also lived alongside of prehistoric creatures, you will never see an evolutionary artist draw or paint a picture of a flamingo wading in a pond next to a brachiosaurus or a possum scurrying out of the way of a T-rex, for it does not fit within the evolutionists' timeline.

FRESH DINOSAUR BONES

Another remarkable fact is that non-fossilized, supposedly 70-million-year-old dinosaur bones were discovered by petroleum geologist Robert Liscomb in north western Alaska in 1961 while mapping for the Shell Oil Company. The bones were fresh, so he assumed that they were recent bison bones. It took 20 years for scientists to uncover the fact that these were the bones of hadrosaurs (duckbills), horned, and large and small carnivorous dinosaurs.

The announcement of these bones was printed in 1985 by the Geological Society of America Abstract Programs.[16] Shortly after this announcement, another story was published in the *Journal of Palaeontology* by Kyle L. Davies.[17] He was amazed that those bones were so remarkably preserved in a relatively fresh state. However, to an unbiased observer, it would seem obvious they were deposited in relatively recent times, perhaps only thousands of years ago, not millions.

Liscomb died in a landslide about a year after discovering the bones. In his honor they named the layer of sediments the Liscomb Bone Bed.

There have been a number of recent stories about bones found in that bed, but evolutionary bias always shows through in the information given and what is left out of the reporting. What should be a prominent part of the reporting on any new information or facts brought to life concerning this deposit, is the fact that these bones for the most part are in non-fossilized condition, a fresh state of preservation. But you will never find this out from reading those stories.

Someone who reviewed this book objected to this information, stating: "The bones are in a fresh state of preservation because they were embedded in the Alaskan tundra—effectively they have been sitting in a freezer for thousands of years." Well, being in a freezer might account for thousands of years of preservation. But these are dinosaur bones (supposedly millions of years old), what kept them fresh during the millions of years before they were placed into the freezer?

This is another perplexing problem for an evolutionist because it does not fit within the evolutionists' timeline. It takes great faith to believe that non-fossilized bones could have lasted 70 million years in a fresh state of preservation without becoming dust.

Stanley Miller and others who devised experiments dealing with abiogenesis did not go out to a forest or jungle and dig here and there and use the material they found for their experiments. They did not go to a junkyard, blindfold themselves, and grope around hoping they would stumble upon junk that would be useful for their experiments, and while still blindfolded use only rocks to assemble what they found. Every experiment carried out is thoroughly and meticulously planned. Nothing is left to chance. Modern scientists who conduct experiments have the luxury of being able to purchase materials, compounds, and chemicals that have already been purified and refined. Glass, metal, plastic, or fuel are never used in raw form; they come from factories.

The laboratories, facilities, and buildings used by these scientists have been planned, constructed, and fabricated.

If evolution did happen, it did not have the amenity and convenience of pure, processed materials and chemical substances ready-made. Miller and others who perform tests always use pre-existing, refined stuff, but the raw materials used by evolution supposedly "popped" into existence. To prove that evolution happened, to get similar results and to do a fair comparison, evolutionists could not use pre-existing materials; their experiments must mimic real life and therefore would have to wait for something to pop into existence again. Even if evolutionists used matter that now exists and were able to make a self-replicating type of life form, their experience would not prove the religion of evolutionism; it would prove it takes intelligence and brainpower to manipulate, control, and engineer the process to create life.

If using pre-existing matter in an attempt to prove evolution, scientists would have to start with a rock and a bit of dust, set it in the sun, sprinkle a little water on it once in a while and let it alone. Then they would have to wait, and wait, and wait... How much time would they have to let pass before they check it for a pulse? I do not know... maybe a million, or a billion, or a trillion years... or until hell freezes over.

LET'S TRY AN EXPERIMENT

Here is another proposal I lay before them, a project they could start today that could be used to confirm their beliefs. First, they need to procure a healthy test subject. Depending on how the organism reproduces—spores, seeds, asexually or sexually—they may need one, a pair, or a group. It makes no difference if it is a fish, a reptile, a mammal, a fowl, a tree, a bacterium, or someone's mother-in-law. Any means possible may be used to get it to evolve, such as heat, cold, drugs, electricity, radiation, selective breeding, different environments, predatory pressures, alchemists, a coven of witches, Gandalf the sorcerer, the alignment of the planets, moonbeams, or harsh language. Using any of the methods listed or any other means put forth, with no time limits or expiration dates, produce a useful feature, organ, characteristic, or attribute in any kind of organism that does not already exist in the chosen species. This would be new information or programming that gives it an evolutionary advantage over its peers.

In the 1995 post-apocalyptic science fiction action film, *Waterworld*, Kevin Costner stars as the nameless mariner, an antihero drifter who sails the shoreless oceans in his trimaran. He is a mutant who has evolved webbed feet and has gills that enable him to breathe underwater. Evolutionists teach that humans and fish have evolved from a common ancestor, and that during one stage in human embryological development the baby has gill slits. So, evolutionary development of gills in humans such as the mariner would supposedly be in the realm of possibility. Because they believe this, he would be the perfect target to shoot for, since they would already be halfway to their goal, the quest to create a new, wholly functioning organ: completely functional gills. They could name this newly evolved human *Humanoid aqueous*. Let us hope they succeed, for the U.S. navy could sure use a man with abilities like that.

Another example would be a gazelle that developed fully functional wings that allow it to fly out of harm's way; that would certainly fit my criteria. Or a tree or bush which grew well-designed legs and feet that enabled it to walk around in order to find better soil or conditions more to its liking—although would a tree that grew legs not also need some kind of receptors in its feet to determine if the soil it had moved to would be a good place to set up housekeeping?

The type of evolutionary proof I am referring to is not the deformed mountain lion[18] that was bagged outside of Preston, Idaho in 2016. This beast had an abnormality never seen before by biologists from the Idaho Fish and Game department. On the left side of the animal's forehead were fully-formed teeth and what appeared to be small whiskers. There were a number of explanations put forth by the department concerning this particular deformity. It could have been the remains of a conjoined twin that died in its mother's womb and was absorbed into the surviving fetus, or it may have been a rare teratoma tumor, which is composed of tissue from which teeth, hair and other body parts can develop. Whatever the case or cause, it was not new information or programming that produced this oddity, but an error or malfunction in the reproduction process of the puma.

Another type of deformity, genetic defect, or mutation that does not establish evolutionary probabilities comes from the two-toed, lobster-claw-footed Wadoma tribe, who live in the Urungwe and Sipolilo districts on the Zambezi river valley in western Zimbabwe. Due to inbreeding brought about by the forbidding of marriage outside of the tribe, one out of four members has a dominant gene that causes "ostrich feet," a condition known as ectrodactyly in which the middle three toes are absent and the two outer ones are turned in. They are sometimes known as the "ostrich footed" tribe. This condition causes some difficulty in walking and running, but does seem to help them while climbing trees. They normally go barefoot, for there are no shoes that would fit them. Whenever needed, custom-made sandals are the order of the day.

Of course, every living thing already has everything it needs to survive. All life forms have been endowed with adaptive capabilities and resources needed to function and thrive within numerous conditions, extremes and environments. Besides wings or gills, it is hard to conceive or imagine what a completely new, fully functioning evolutionary advantage feature, ability, or organ would possibly look like. Evolution had no oversight—no one to gather or mix the chemicals, tweak the amounts, or fine-tune or adjust the timing. The evolutionist might complain that the kind of experiments I recommend would not be a fair comparison, for the early earth had an atypical environment; conditions were dissimilar and the atmosphere was very different. As we have already discussed in chapter two, of course the atmosphere was vastly different, there was no oxygen; oxygen would cause any newly formed compounds to oxidize and break down before they could be of use to a self-forming cell. Thus any life would have had to be an anaerobic life-form in that hostile oxygen-free environment. However, this problem is seldom discussed by evolutionists—the fascinating story of how simple protozoan or bacterial anaerobic lifeforms evolved into more complex oxygen-loving organisms. This would have certainly made a great story for one of Paul Harvey's famous *The Rest of the Story* segments which always concluded with his tag line, "And now you know the rest of the story." Of course there is no evidence for this, for no matter how far back in time evolutionists go, they find no evidence of simple kinds of life. Any so-called ancient life has always been shown to be as complicated and complex as any known living organism.

INDEX FOSSILS

Let's finish out this chapter with a discussion of another method of dating fossils: The *index fossil*, also called a *zone*, *indicator*, or *guide* fossil. These fossils are used to help define and identify geologic periods or faunal stages.

The *American Heritage Science Dictionary* 2002 reprints Houghton Mifflin's definition of an index fossil:

> The fossil remains of an organism that lived in a particular geologic age, used to identify or date the rock or rock layer in which it is found. The best type of index fossils are usually those of swimming or floating organisms that evolved quickly (and therefore did not cover a long span of geologic history) and were able to spread over large areas. Ammonites and graptolites are good index fossils.

Index fossils are from creatures assumed to have lived in a particular geologic age. They are characterized as types of creatures that were able to spread and disperse over large or broad areas of the earth's surface. They were rapidly evolving and thus were short-lived (in geological terms); they existed for a brief period, so any rock layer containing that fossil can be linked to a specific geologic time.

Not all fossils are used for dating. The horseshoe crab has supposedly existed for over 400 million years. Its fossil remains have been found in many rock strata, and it is still alive today. It is not useful in pinpointing a geological era. It covers too much historical ground.

Commonly used index fossils are ammonites, graptolites, brachiopods, nanofossils, and trilobites. The nanofossils are abundant microscopic remains of calcareous nannoplankto and coccolithophores from various eras. Nanofossils are the primary method of dating marine sediments.

Read the Wikipedia article about the ammonite:

> Starting from the mid-Devonian, ammonoids were extremely abundant, especially as ammonites during the Mesozoic era. Many genera evolved and ran their course quickly, becoming extinct in a few million years. Due to their rapid evolution and widespread distribution, ammonoids are used by geologists and paleontologists for biostratigraphy. They are excellent index fossils, and it is often possible to link the rock layer in which they are found to specific geological time periods.[19]

This information applies to all index fossils. However, how do they know index fossils "ran their course quickly" and only existed for a brief period of time? Consider the numerous creatures discussed above that were thought to be dead and gone, only to reappear millions of years later. How can they make that kind of assumption, when the methods used to place index fossils in the geologic column are arbitrary and indiscriminate? Most importantly, how do they know these organisms evolved?

Have you grasped and become conscious of the fact that it is belief that fires the engine of evolutionism and its geological time column? Do you understand that it is just an opinion and faith in a religion, not pure science which assigns a fossil's place in the column? Do you realize that not a single scientific test is used to corroborate, validate, or determine the era, epoch, or time period assigned to any index fossil?

Sometimes you will read about radiometric testing being done on fossils. However, the results are already predetermined, for all test results must conform to the dates and parameters established by the geological column. For someone thinking logically, this information should be a wake-up call, but because of the evolutionists' blind faith, they remain oblivious and continue to ignore these facts.

NOTES ON CHAPTER 10

1. "The Earth Forms," *BBC Science* web site, n.d., http://www. bbc. co. uk/ science/earth/earth_timeline/earth formed.
2. "How Old Is the Earth," Extreme Science web site, n.d., http://www.extremescience.com/earth.htm.
3. "Geochronology," Wikipedia web site, August 28, 2017, https:// en. wikipedia.org/wiki/Geochronology; "Radiometric Dating," Wikipedia website, September 23, 2017, https://en.wikipedia. org/ wiki/ Radiometric _dating.
4. "Do You Believe Radiometric Dating Is an Accurate Way to Date the Earth?" UCSB ScienceLine web site, April 3, 2012, http:// scienceline. ucsb.edu/getkey.php?key=2901, paras. 1–2, 4, 6–7, 16.
5. Ibid., paras. 11, 15–16.
6. Brian Thomas, *Dinosaurs and the Bible* (Eugene, OR: Harvest House Publishers, 2013), pp. 72, 73.
7. Arielle Duhaime-Ross, "Ancient Comb Jellies Had Skeletons, but They Still Lost the Arms Race," The Verge web site, July 10, 2015, https:// www.theverge.com/2015/7/10/8929251/comb-jellies-skeletons; Laura Geggel, "Rare Fossils of 400-Million-Year-Old Sea Creatures Uncovered, LiveScience web site, July 8, 2015, https:// www. livescience.com/51474-morocco-marine-fossils uncovered.html.
8. R. K. Sepetjian, "The Geologic Column: Invented to 'Free the Science from Moses,'"Across the Fruited Plain web site, October 11, 2011, https://sepetjian.wordpress.com/2011/10/11/the-geologic-columninvented-to-%E2%80%9Cfree-the-science-from-moses%E2%80%9D/, para. 8.
9. The standard geologic column supposed by evolutionists does not exist to any substantive extent; is found in only one place, textbooks. This column if complete would be an estimated 100 to 200 miles thick. What little they do find is made up of layers of sedimentary rock that contains fossils, to which evolutionists assign dates based solely upon the assumption of evolution with its accompanying long ages of time.
10. Michael Benton, "Accuracy of Fossils and Dating Methods," Actionbioscience web site, January 2001, http://www. actionbioscience. org/evolution/benton.html, para. 3.

11. Joseph McCabe, "Alexander von Humboldt," *A Biographical Dictionary of Modern Rationalists* (London: Watts, 1920), reprinted on the Freedom from Religion Foundation web site, https://ffrf.org/ news/day/ dayitems/ item/14545-alexander-von-humboldt
12. "The Fossil Record," Louisiana Department of Natural Resources web site, n.d., http://www.dnr.louisiana.gov/assets/TAD/education/ BGBB/ 1/ fos_measuring.html, para. 4.
13. Dave E. Matson, "How Good Are Those Young-Earth Arguments?" TalkOrigins Archive web site, February 12, 1995, http://www. talkorigins.org/faqs/hovind/howgood.html
14. Tom DeRosa is Executive Director and Founder of *Creation Studies Institute*. He taught Chemistry and Physics in public and parochial schools for over 30 years. Has a strong academic background in Chemistry and holds a Master's degree in Education from Florida Atlantic University. He has led teaching seminars and has been recognized by the Broward County School Board for his expertise in science education. Has contributed to several published science-teaching manuals and was the Director of the Regional Science Talent Search for high school students. And among other duties served as Director on the Florida Academy of Science Executive Board. "Meet the Staff," Creation Studies Institute web site, n.d., http://www.creationstudies.org/about-us/csi-staff
15. Carl Werner, *Living Fossils—Evolution: The Grand Experiment* (Green Forest, AR: New Leaf, 2009).
16. An initial announcement was printed in 1985 in the *Geological Society of America Abstract Programs* vol. 17, p. 548.
17. Kyle Davies, "Duckbill Dinosaurs (Hadrosauridae, Orinithischia) from the North Slope of Alaska," *Journal of Palaeontology* 61(1), pp. 198–200.
18. Nina Golgowski, "Mountain Lion Found with Teeth Growing Out of Its Head," *The Huffington Post* web site, January 10, 2016, http:// www.huffingtonpost. com/entry/mountain-lion-teeth_us_56925 ef0 e4b0c 8beacf76153.
19. "Ammonoidea," Wikipedia web site, September 24, 2017, https://en. wikipedia.org/wiki/Ammonoidea, Distribution, para. 1.

> One must conclude that, contrary to the established and current wisdom a scenario describing the genesis of life on earth by chance and natural causes which can be accepted on the basis of fact and not faith has not yet been written.*

CHAPTER 11

EXTRA-LARGE CRITTERS

IN THESE NEXT FOUR chapters we will reflect upon the climatic conditions of the antediluvian world, consider its landscapes, flora, fauna, and people, contrasted with our present world. We will explore the reasons for these differences and why they were possible. We will also contemplate the reasons, causes and problems concerning the extinction of the enormous beasts, fowl, and insects that no longer inhabit this biosphere.

When researchers examine the effects left behind from ancient creatures, such as skin impressions found in stone, tracks, and fossils; comb through coprolites to identify what they ate, they find that their environment was in some ways similar to what we find today in numerous places and environs across the globe. However, to allow for the great excess of vegetation pole to pole and the immense proportions attained by animal life, something different governed and controlled the climates, environment, and atmospheric operational systems.

In God's original creation, as evidenced by the fossils of plants entombed in the rocks, there was an astonishing variety of lush and flourishing vegetation. For example, fossils of club moss have been found

---

*Hubert P. Yockey (Army Pulse Radiation Facility, Aberdeen Proving Ground, Maryland, USA), 'A calculation of the probability of spontaneous biogenesis by information theory'. *Journal of Theoretical Biology*, vol. 67, 1977, p. 396.

that were 150 ft tall, whereas today a foot and a half is all they can manage. Immense deposits of coal and huge oil pools have been found over the entire earth, denoting a much greater mass and diversity of vegetation, making the whole earth a veritable paradise. This certainly fits with God's description of His creation when the text states: *"And God saw every thing that he had made, and, behold, it was very good"* (Genesis 1:31).

Today about 71 percent of the Earth's surface is water-covered. However, there is much evidence that in the past the oceans were probably smaller. One-time Professor of Hydraulic Engineering, Henry M. Morris wrote:

> There is also much evidence that sea level was once much lower relative to the land surfaces than it is at present, implying either that the amount of water in the ocean was much smaller, or that some parts of the sea bottom have dropped, or both.[1]

Smaller oceans and more land surface would also help account for the vast amount of stored vegetation found in huge seams we call coal. The natural environmental and ecosystems of the original creation seemed similar, but operated in dissimilar ways. Something was certainly different in a number of important respects, and whatever that was allowed vegetation such as trees to grow at both the North and South Poles.

One difference was the hydrological system God designed for precipitation. It never rained, *"But there went up a mist from the earth and watered the whole face of the ground"* (Genesis 2:6). Everything probably grew like weeds, for the air had a greater concentration of carbon dioxide, and the soil contained all the proper amounts of trace minerals and nutrients needed for optimal plant growth. However, that does not explain how the land at the poles could be totally ice-free and grow thriving temperate forests long ago, when it now stays dark for months at a time, gets intensely cold, and is covered by glaciers and ice.

Evolutionists believe the poles were ice-free, for the earth had a much warmer climate in the past, but evolutionism tends to gloss over this problem, for it has no good explanation for such a phenomenon. George McCready Price said this about the earth's past climate: "But it is needless to go through the systems one after another, for they "uniformly testify that a warm climate has in former times prevailed over the whole globe."[2] Another outstanding authority, Professor Alfred R. Wallace, stated:

> There is but one climate known to the ancient fossil world as revealed by the plants and animals entombed in the rocks, and the climate was a mantle of spring-like loveliness which seems to have prevailed continuously over the whole globe. Just how the world could have thus been warmed all over may be a matter of conjecture; that it was so warmed effectively and continuously is a matter of fact."[3]

Noted geologist Sir Henry H. Howorth wrote:

> The flora and fauna are virtually the only thermometer with which we can test the climate of any past period. Other evidence is always sophisticated by the fact that we may be attributing to climate what is due to other causes. But the biological evidence is unmistakable; cold-blooded reptiles cannot live in icy water; semitropical plants, or plants whose habitat is the temperate zone, cannot ripen their seeds and sow themselves under arctic conditions."[4]

Although these quotes from Price, Wallace, and Howorth are well over half a century old, this information, as shown by recent studies, has been continually confirmed since that time by paleontologists and biologists. For example, consider these two recent stories. Writing for the LiveScience web site, Katia Moskvitch stated the following in her March 24, 2014 article "Dinosaur Era Had 5 Times Today's CO2":

> Dinosaurs that roamed the Earth 250 million years ago knew a world with five times more carbon dioxide than is present

on Earth today, researchers say... Huge amounts of this greenhouse gas made the climate during the Jurassic Period extremely humid and warm, said geoscientist Douwe van der Meer, lead author of the study and a researcher at Utrecht University in the Netherlands... "The higher CO2 levels [must] have [had] significant effects on the planet's climate, and its flora and fauna," he said.[5]

A February 28, 2012 article in the *Daily Galaxy* via Geology and The Daily Galaxy web site, "EcoAlert: Earth was Stifling Hot During Peak Age of Dinosaurs," stated that:

> ...High temperatures and possibly more atmospheric carbon dioxide caused forests to extend much closer to the poles and grow almost twice as fast as they do today... "Some fossil trees from Antarctica had rings more than two millimeters wide on average. Such a rate of growth is usually only seen in trees growing in temperate climates. It tells us that, during the age of the dinosaurs, polar regions had a climate similar to Britain today," explains co-author Dr. Howard Falcon-Lang. "Our research shows that weird monkey puzzle forests covered most of the planet, especially in the steamy tropics. At mid-latitudes there were dry cypress woodlands, and near the North Pole it was mostly pines," said Emiliano Peralta-Medina, who led the study...[6]

The *global warming* crowd seems to be composed largely of evolutionists, tree-huggers, and similar characters. Those folks are afraid the climate is getting warmer and will melt the icecaps in Greenland and the polar icecaps as well. This would cause most of the world to flood. Their fears were conveyed in the 1995 post-apocalyptic science fiction action film *Waterworld*, where the ice caps melted and the world became a shoreless ocean (spoiler, there was still a patch of dry land). The water issue is a problem, a conundrum for that bunch of fatuous and ill-advised folks.

These questions should help clarify the matter.

Since there were no icecaps or glaciers to melt in the days of the megafauna, T rex, Stegosaurus, and the rest of the reptile gang, why was the world not flooded? Why were the oceans smaller and the climate much warmer in those prehistoric days? Where did all the water for the present day oceans come from (which covers three-fifths of the earth), and the water for the glaciers and icecaps, which they think may melt and overwhelm the land in the near future?

WHO DELIVERED THE WATER?

Although some evolutionists rarely discuss, tackle, or even mention the issue, many astronomers do address the question of how the original amount of H_2O was acquired and unceremoniously dumped on this planet. It had to come from somewhere other than the earth, for they are sure that water did not exist on our planet directly after its formation.

On NASA's website they suspect that comets brought water to the earth:

> But if life formed so quickly on Earth and there was little in the way of water and carbon-based molecules on the Earth's surface, then how were these building blocks of life delivered to the Earth's surface so quickly? The answer may involve the collision of comets with the Earth, since comets contain abundant supplies of both water and carbon-based molecules. [...]
>
> Since their orbital paths often cross that of the Earth, cometary collisions with the Earth have occurred in the past and additional collisions are forthcoming.[7]

In NASA's tale, they reason that comets and asteroids travel in their own orbits or paths around the sun, and once or twice a year earth passes through this swarm of space debris. This would be like an airplane flying through a large flock of ducks—it is bound to hit at least a few. Over the course of many millennia, earth's infrequent collisions with dirty ice-balls brought enough water to fill the oceans.

In 2013, the *Smithsonian Magazine*[8] tackled the question that has been hotly debated by astronomers for decades, which two particular denizens is the galactic water truck responsible for bringing the needed liquid refreshment to our world? Is it comets that utilize the Solar Express System, or is it asteroids that bear the burden for on-time water delivery? We may never know, for as stated in the story: "Adjudication between the two is a challenge, and over the years *scientific judgment* has swung from one to the other." (emphasis added)

The reporter calls it "scientific judgment," as if it was real science, in which they could actually perform experiments or test their options concerning ancient history. In reality, it is probably just two guys, who once each decade flip a coin, best two out of three, to determine which delivery system was used—heads it was asteroids, tails it was comets. In other words, they are not sure which leftover materials from the formation of our Solar System, those celestial bodies that orbit our Sun, the gravel, stones, boulders, and sarsen strewn between the planets, was tasked with watering planet Earth.

Another question regarding these celestial pebbles flying around in space is, did all of them contain reservoirs of water? Who or what filled each cistern in the beginning? Is not it possible that many of them could have been dry, devoid of that precious commodity? Also, during the millions of years they spent just lounging around in space, what kept their water from being vaporized into the vacuum of space? My best guess would be zip-lock plastic bags.

So, assuming most comets have reservoirs of water, let us explore a little further, ruminate just a bit on evolutionism's beliefs and ideas and see if the comet delivery system can pass the logic test. Can they really explain why the Earth was not flooded 65 million years ago when there were no polar icecaps to melt, and the oceans were smaller?

Consider the vastness of space and the immense distances between the stars, planets, and galaxies. According to the present consensus on how the universe operates, nothing is stationary; everything is moving rapidly through space. This includes the relatively tiny speck of rock, the planet we call home, which supposedly spins at 1,037 mph and moves at 66,666 mph in orbit around the sun—much faster than a speeding bullet (an average bullet travels at 2,500 feet per second, or about 1,700 mph). The sun, relative to the Milky Way moves through the galaxy at 45,000-52,000 mph or 12 miles per second. The Milky Way along with the rest of our local group of galaxies is moving through space at 1.2 million mph (with all this movement, it makes one wonder, why don't we feel any of it?).

Our Earth is struck every day with debris from space, and once in a while a sizeable chunk of rock hits our planet. However, to fill our oceans with water a much more efficient, consistent means must have occurred, for no half-hearted effort would do. As we reflect upon that information, we come to the realization that any complex targeting system designed by Milky Way Inc. (MWI) for its galactic watering trucks would be the envy of all space and military engineers, for it is extremely accurate and, regardless of the distance, has the ability to lock upon and track numerous objects at the same time.

In order to achieve the goal of filling earth's empty basins with water to create oceans, each day for millions of years, this ancient business plucked asteroids and comets from the empty reaches of space to use as delivery vehicles. Their main problem was getting each tiny ice-laden rock up to speed, for their target was moving at speeds in excess of 66,000 mph. They solved this dilemma by outfitting each fragment of rubble with its own course-correcting, propulsion system. It was then programed with precise instructions and directions how to get to the very tiny, quickly moving chunk of rock, our earth.

MWI did an extraordinary job, for getting tiny fragments of rock to consistently hit such a swift-moving target is akin to hitting a bullet with another bullet. To do that unswervingly over millions of years with innumerable pieces of rock—just what are the odds? MWI was able to accomplish its impossible mission, for 65 million years ago the oceans were filled with the required amount of liquid, just in time for the pyrotechnic display caused by the asteroid that wiped out the dinosaurs and most other life forms on planet earth.

From that day to this, they have continued to deliver water via the comet/asteroid system, so that the great increase in the volume of water present today, over the much smaller amount sloshing about in the oceans during the time of the dinosaurs, is all accounted for. If MWI continues its exemplary on-time water shipments for the next 65 million years, all that additional fluid added to the already overflowing oceans, will truly make treading water a necessary survival skill for all terrestrial creatures.

After evolutionists spent many years trying to develop a workable idea of where earth's water came from, the best guess, the finest idea they could conceive or envision, is the childish belief of special delivery by the *Space Debris Fairy*. Do not hold your breath waiting for a plausible, credible, or believable solution. Evolutionists have to resort to silly ideas, for they reject the obvious: The only credible explanation or answer for the existence of water on earth is that God created it. Surprisingly, this earth has been totally submerged with water twice in its history. The flood during the time of Noah was the second occurrence of total inundation. The first time was during the first two days of Earth's creation. It was not until the third day that dry land appeared: "*And God said, Let the waters under the heaven be gathered together unto one place, and let the dry land appear . . . And the evening and the morning were the third day*" (Genesis 1:9–13).

Evolutionists' line of reasoning also showcases the ignorance of the inevitable question whenever the flood of Noah is mentioned in some news story. Many mockingly ask, where did all the water go if the whole world was once covered with flood waters? This question truly shows either they are willingly ignorant or sincerely oblivious to their lack of skill concerning reasoning, analysis, and deduction. Within their own belief—the fear that the polar icecaps might melt and inundate coastal regions, making cities such as Boston, New York, and Miami uninhabitable—lays one answer. If there is enough ice to melt to flood the earth, the water did not go anywhere; it is still here, in the oceans that now cover 71 percent of the earth, and the rest is locked away in the polar icecaps! Of course, the unenlightened, in-the-dark evolutionist might reply, if all the icecaps did melt, there still is not enough water to cover most of the mountains, especially peaks as high as Mt. Everest.

The answer is not hidden, but lies in plain sight. Even the most insensible evolutionist knows Everest, like all mountain ranges, is covered with thousands of feet of sedimentary rock that contain fossils of sea life. So it is obvious that if Everest was once covered with sea water; every other mountain range would also have been submerged, along with every bit of dry land. We will cover this bit of obliviousness in greater detail shortly.

BEREFT OF THE GIANTS

Leaving behind the evolutionists' space debris and lack of water nonsense, aside from the landscape, with its worldwide, pole-to-pole abundance of vegetation, there was something else very impressive about the original world God created: the size of the animals, fowls, and insects. Many of the largest creatures to ever live are now extinct. The earth is now bereft of many of the most impressive beasts that once flew the skies, swam the seas, and prowled this planet. Professor Wallace was quite correct when he stated: "It is quite clear, therefore, that we live in a zoologically impoverished world, from which all the hugest and fiercest and strangest forms have disappeared."[9]

Let us now consider some sizable, but relatively unknown creatures that meandered around the ancient world. Most people are familiar with the magnificent reptiles called dinosaurs, but not with the largest land invertebrate ever: *Arthropleura*, a centipede-like creature that grew to over 9 ft in length. If those creatures were still around today and your house became infested with them, for a creepy-crawly this size you could not call a pest-control company; it would take heavy weapons processed by the military to dispose of this beast.

Nor are most folks familiar with the giant *marsupial wombats* with backward-facing belly pouches large enough to carry an adult human, or the recently discovered *coal turtle* that grew to the size of a small car. I wonder how much turtle soup could be made from a reptile that size? For crustacean lovers who want to gorge on escargot, fossil shells have been found measuring nearly a foot in diameter. If you feel charitable and want to treat the entire neighborhood to a lobster feast, fossils belonging to that family have been found that measured at least 6 ft in length. And after you have found an industrial-sized pot, filled it with water, and brought it to a boil, good luck with your struggle to get a live one that size into it.

If amphibian legs are more to your taste, and you happen to be in Cameroon or Equatorial Guinea, you might try your hand down at the local pond stalking the world's largest living frog, the Goliath, which can grow up to 13" long and weigh a hefty 8 lbs. However, if you were hunting ancient croakers, you would need to be careful in order to avoid being eaten yourself. For pursuing those amphibians, forget the BB gun or air rifle and take something with a little more firepower, say an AK-47 Russian assault rifle, along with a few concussion grenades for back-up, because your prey would measure 6 to 10 ft long with a head some 20" in length.

Until fairly recently, about 500 years ago, if you wanted to make an omelette from just one egg that could feed a large family for breakfast, the pale green, 10" long, 7" wide egg of a *moa* would have been just right. Supposedly in existence for 100 million years and contemporary with dinosaurs, it was a 500 lb flightless bird that grew to 12 or 13 ft. It lived in the forest fringes and savannahs of New Zealand and surrounding islands until hunted to extinction by the Maori in the 1400s. There was another species of giant bird, allegedly in existence for 60 million years, known as the *elephant bird*, which roamed the same islands as the moas as recently as 350 years ago. Although not as tall, they were massive, the largest birds that have ever lived. Four species of these birds are believed to have existed, but became extinct sometime during the mid-seventeenth century—about the same time firearms were introduced to Madagascar.

These flightless, ostrich-like birds grew up to 10 ft tall and weighed about half a ton. They laid 22 lb eggs, which were well over a foot long and 8" in diameter—that's over 3 ft in circumference! Its volume would be the equivalent of close to 200 chicken eggs, enough to make omelets for over 100 people!

As chicken eggs come in various grades, I wonder, did these eggs also come in small humongous, medium humongous, large humongous, and extra-large humongous grades? Think of the problem eggs that size might have caused when buying weekly groceries for a very large family. After purchasing a dozen of those eggs at the grocery store, loading and unloading the egg carton from the car would require a forklift. For dinner, what a feast a bird that huge would provide, enough food for an entire small village, or, if you married a Filipino, for your wife's extended family. But where would you find cooking instructions for a half-ton ostrich-like fowl? How many hours, probably days, would it take to cook? After preparing the bird, filling it with stuffing (would 300 lbs be sufficient?), and pre-heating your oven to 325 degrees, the biggest mystery, how would you cram something that size into your oven?

For ornithologists who prefer to view aves on the wing, soaring through the skies, the Giant Teratorn (*Argentavis magnificens*), discovered in Argentina in 2006 would have been a breathtaking sight. Weighing between 140 and 180 lbs, with a wingspan of 20 to 26 ft, this eagle would have been a sight any bird-lover would treasure for years to come. However, there is something I find very frightening about a raptor of these proportions: an eagle's killing power is way out of proportion to its size. Though jackrabbits and other small mammals are its main prey, eagles do attack and can kill game much larger than themselves. Small deer, goats, and even bear cubs are sometimes on the menu of large eagles. They have been seen killing wolves and fighting off bears and coyotes in defense of their young and prey, and their powerful wings allow them to carry off animals almost half their body weight. Throughout the world, their hunting prowess and ferocity inspire reverence and fear.

An eagle's attack is silent, swift, and lethal. By the time its quarry sees its shadow, it is too late. Swooping in on prey with speeds in excess of 100 mph, its deadly talons and powerful beak make short work of the unlucky victim. For an eagle the size of the Giant Teratorn, anything smaller than a full-grown elephant or hippo could be on the *carte du jour*, including inquisitive bird-watchers.

In the skies, dodging those birds flew the *Meganeura*, the largest flying predatory insect in the history of the earth. A flyswatter-proof, dragonfly-like insect with a 38" wingspan that grew as long as the height of the average toddler. Other creatures that flew alongside these insects were the flying winged lizards, *pterosaurs*, reptile-like, with fully toothed jaws and long tails. One in particular grew to an amazing size, the giant *Quetzalcoatlus*, which had a 30–40 foot wingspan and when fully mature would have weighed in at about 450 lbs.

Here in the Philippines, cockroaches can grow over 2" long, and they fly! I am very grateful they do not grow as large as those that inhabited the ancient landscape, for those that scurried among the forest litter in the dark and flew through the brush of the jungles grew to the size of house cats. Imagine traveling on your motorcycle, helmet on for safety, zipping along at 60 mph on some winding picturesque jungle road, just enjoying the scenery, and that sturdy, armor-shelled, feline-sized creature struck you in the chest. Its mass would knock you off the cycle, spinning head over heels; you would hit the pavement with a thud and skid for a short distance. Then, after some unplanned gymnastic tumbling left you hurting, dazed, and very stunned, it would fly away like nothing happened. Was it just your imagination, maybe a dream, or hallucination, but did you actually hear faint laughter, and a wee small voice saying... "Heeey, I just zappped mee awe nother one!"

Also scampering around on the forest floor was the heart-attack-inducing *pulmonoscorpius*, a scorpion very similar to today's varieties but which could grow to over 3 ft in length. If slumbering, it was something you would not want to disturb by stepping on it, for it was armed with large sharp claws and a venomous stinger. If you did accidently disturb a nest of those frisky land-loving arthropods and were running for your life, diving into the nearest body of fresh water to try and effect an escape might not be a good idea. For lurking beneath the surface of many rivers and lakes was the monster-movie-sized *Jaekelopterus rhenaniae*, a giant sea scorpion, an ugly-looking, 8-ft-long, lobster-like beast with claws the size of a grown man's head.

There were giant sloths with massive skeletal structures that grew to more than 5 tons in weight and 19 ft in length, which were able to reach as high as 17 ft and were taller than a bull African Bush Elephant. I wonder, as modern sloths only defecate once a week and can store up to a third of their body weight in feces, was this habit passed down from this ancient beast? If so, what a humungous pile of excrement those beasts would have produced! Talk about being "full of it." Paul Gautschi would have had stars in his eyes, thinking of all the free compost material, the brown gold those beasts would have produced for his garden.

Baluchitherium
Mark Johnson

There were beavers that were 8 ft in length and weighed 450 lbs—a lot larger than the average adult black bear today. Imagine a trapper working his trap-line and coming across a beaver that size! A normal beaver trap would not suffice—a bear trap would be required. But what a bounty pelts of that size would bring! For deer hunters looking for a trophy animal, the *Irish elk* was one they would have prized, for it was one of the largest deer to ever live. Its range extended across Europe, northern Asia, into parts of China and Africa. It stood about 7 ft tall at the shoulders and carried the largest antlers of any known Cervidae (deer), about 90 lbs., with a spread of 12 ft from tip to tip. The largest specimens weighed in at over 1,500 lbs, similar in weight to the Alaskan Moose. There were a host of other large mammals that rivaled in size some of the largest dinosaurs. One such rival was the *Baluchitherium*, believed to be the largest terrestrial mammal of all time. Although it was hornless and had a neck longer than modern types, it clearly possessed other features that indicate it was a rhinoceros. It stood 18 ft at the shoulder (the average height of a bull giraffe is 18 ft) and weighed approximately 20 tons, nearly twice the size of the largest elephant ever recorded. A 6-ft-tall man standing directly under this beast would have to reach up to tickle its belly.

In the seas swam the fearsome shark called the *Megalodon*, whose name in Greek means "big tooth." It reached an estimated length of 60–100 ft and, amazingly, might have weighed 100 tons (an adult T rex weighed in at a measly 8–9 tons). The Megalodon was probably the biggest predatory marine or land creature in the history of the planet. It also had serrated teeth, of which the largest found so far are over 7" long. Compare that with the teeth of a modern Great White shark, whose teeth only grow to about 2" in length. The oceans of the past must have been teeming, swarming with life, for think of the abundance of prey needed to keep such a massive fish fed.

There were giant freshwater crocodiles, like *Sarcosuchus imperator*, whose heads were over 5 ft long; these animals grew to over 40 ft and weighed almost 9 tons. For bite-size snacks before its main meal, there were the *Megaprianha paranensis*, 3-ft-long piranhas to dine upon. For those who have *ophidiophobia*, the fear of snakes, the giant slithering reptiles which lurked in the ancient jungles and waterways would have given them a real reason for fear. One such monster was called the *titanic boa*, a type of constrictor like an anaconda or boa. It weighed in at over a ton and grew to over 40 ft in length. The jungles were the home territory of *Gigantopithecus*, a gigantic ape that, though not quite as big as the fictional King Kong, was nonetheless still imposing, for he stood 10 ft tall and weighed a hefty 1,200 lbs. Who knows, maybe his home turf was on the real Skull Island. (Writing about Kong gives me something else to ponder and reflect upon. It is a puzzle to me. In the movie *King Kong*, why would the natives on Skull Island build a huge stone wall to keep Kong out, and then make a wooden door big enough for him to fit through?)

Roaming the antediluvian landscapes with all the other titans was *Arctodus simus*, a 12-ft-tall, 1-ton, giant short-faced bear, as well as the fearsome *Pachycrocouta brevirostris*, a hyena with an ear-splitting cackling laughter, for it was a huge beast, double the size of its modern relatives. Are you familiar with cute little armadillos that are native to the Americas? Most are small, but there is one modern armadillo that grows up to 120 lbs. However, it is dwarfed by its ancient relative. Along with all the other oversized creatures, this kind also came in an extra-large variety, about 200 lbs. worth. Do not try to kill one of these armored beasts with your car; you will only make it mad.

In our cursory survey of the ancient *megafauna* (Greek *megas* "large" and Latin *fauna* "animal"), it seems that in most cases, the terrestrial creatures that walked the land, flew in the skies, and swam in the water were more impressive, more abundant, robust, bulkier, and must have been on huge doses of steroids, for in many cases they were not just a little pumped-up and oversized, but they grew to gigantic proportions unheard of in today's world. For evolutionists, this information should be somewhat puzzling, an enigma. But many, it seems, are not cognizant of the fact, nor acknowledge the truth, that it goes against the grain, the idea that things have been *progressing*, that life is always moving up the phylogenetic tree, as evolutionists envision. Now, being larger is not necessarily better, there may be some drawbacks, such as more food being required. This idea could be debated. However, this is what we have seen from the evidence, the flora and fauna were notably larger.

MODERN ANIMAL GIANTS

Let us now consider a few of the largest creatures, aquatic, land-dwelling, and flying fowl that now occupy our world.

The blue whale is probably the largest animal to ever inhabit the earth. Surprisingly, they are warm-blooded mammals. They do not have gills like fish, but are equipped with lungs and can hold their breath for 20 minutes or longer, but must come to the surface to breathe air. They have been measured at 111 ft long—17 ft longer than the length of a basketball court—and weigh up to 160 tons (320,000 lbs.). Their hearts, which are the largest of any mammal, can weigh up to 1,300 lbs and are the size of a Volkswagen Beetle. This largest of creatures feeds on some of the smallest, krill, shrimp-like crustaceans that it strains through baleen plates to filter out food from the water. It needs about 8 tons each day to satisfy its hunger. Unlike most land animals, the blue whale does not rely on its skeleton to support its huge body. Instead it uses the buoyancy of water, the phenomenon called *zero gravity*. We have all experienced that feeling of being weightless whenever we float in a swimming pool: The buoyancy of water helps to keep its body afloat.

The largest land animal is the African elephant. The largest elephant ever recorded was shot in Angola in 1956. It had a shoulder height of 13 ft, which is about a yard taller than the average male African elephant, and it weighed about 24,000 lbs (12 tons). Elephants can move both backwards and forwards. They swim well, but cannot jump, trot, or gallop. They have two gaits: A walk and a faster gait that is comparable to running. Because of their bulk they cannot tolerate falls, and even a slight stumble could break bones or damage enough tissue to have terminal consequences, so they spend most of their lives plodding along carefully. This is why a relatively small moat, in which a healthy man with a running start could easily jump over, is employed in many modern zoos to keep elephants confined.

The wandering albatross is an amazing and majestic bird. At about 12 ft, it has the largest wingspan of any living bird. There are also accounts of albatrosses with wingspans as great as 17 ft across. Their long wings allow these birds to glide effortlessly over the ocean for hours at a time without flapping their wings. They only come ashore every other year to reproduce. The three largest birds by weight, in order, are the Andean Condor from South America (33 lbs), the Mute Swan of Europe and Western Asia (39 lbs), and the heavyweight champion, the Kori Bustard from the southern savanna areas of Africa takes the prize at 41 lbs!

All of these are impressive creatures, wondrous to behold. However, with the lone exception of the blue whale, our modern landscape, contrasted with the old, is inhabited only by pygmies, small breeds, puny replicas, and much smaller duplications of the past glories, grandeur, and opulence of the animal kingdom. People are fascinated by the enormous creatures that used to roam this earth. Many ask the question "Why don't we see insects, birds or animals today the size they grew too in the past? They were massive in the past, so why not today?" After all, they have had 65 million years to re-evolve, recuperate, and recover their former splendor and majesty.

Before we consider those things, in the next chapter we will explore a little further the humans that inhabited the antediluvian world, the conditions they lived under, and what their civilization might have been like.

NOTES ON CHAPTER 11

1. John Whitcomb and Henry Morris, *The Genesis Flood* (Grand Rapids, MI: Baker Book House, 1961), p. 124.
2. George McCready Price, quoted in Alfred Rehwinkel, *The Flood in the Light of The Bible: Geology and Archaeology* (St Louis, MO: Concordia, 1951), p. 8.
3. Alfred Wallace, quoted in Jerry Blount, *Noah and the Great Flood: Proof and Effects* (Columbus, OH: Gatekeeper Press, 2017).
4. Henry Howorth, quoted in Rehwinkel, 1951, p.7.
5. Katia Moskvitch, "Dinosaur Era had 5 Times Today's CO2," LiveScience web site, March 24, 2014, https://www.livescience.com/44330-jurassic-dinosaur-carbon-dioxide.html, paras. 1, 4.
6. "EcoAlert: Earth was Stifling Hot During Peak Age of Dinosaurs," The Daily Galaxy web site, February 28, 2012, http://www.dailygalaxy.com/my_weblog/2012/02/ecoalert-earth-was-stifling-hot-during-peak-age-of-dinosaurs-1.html,paras 1–2.
7. Don Yeomans, "Why Study Comets?" NASA web site, April 1998, https://stardust.jpl.nasa.gov/comets/comets1.html, paras. 1–2.
8. Brian Greene, "How Did Water Come to Earth?" *Smithsonian Magazine*, May 2013, http://www.smithsonianmag.com/science-nature/how-did-water-come-to-earth-72037248/
9. Wallace, quoted in Blount, 2017, p. 21.

> I have said for years that speculations about the origin of life lead to no useful purpose as even the simplest living system is far too complex to be understood in terms of the extremely primitive chemistry scientists have used in their attempts to explain the unexplainable that happened billions of years ago. God cannot be explained away by such naïve thoughts.*

CHAPTER 12
GOLIATHS AND PRODIGIES

AN AMAZING WORLD, WAS it not? But before we ponder the antediluvian society that dwelt in harmony with the abundant flora and colossal fauna which once lay claim to our planet, consider the reason God created this earth: It was to be man's home and habitation. *"For thus saith the LORD that created the heavens; God himself that formed the earth and made it...he formed it to be inhabited..."* (Isaiah 45:18). Man was to be the King of this planet, for he was given dominion over everything:

> And God said...let them have dominion over...all the earth, and over every creeping thing that creepeth upon the earth...and subdue it: and have dominion over the fish of the sea, and over the fowl of the air, and over every living thing that moveth upon the earth. (Genesis 1:26, 28)

What is man, that thou art mindful of him? And the son

---
*Ernst Chain (world famous biochemist), as quoted by R.W. Clark, in the *Life of Ernst Chain: Penicillin and Beyond*, Weidenfeld & Nicolson, London, 1985, p. 148.

of man, that thou visitest him … Thou madest him to have dominion over the works of thy hands; thou hast put all things under his feet. (Psalms 8:4–6)

For those not aware of the true history of this planet, people and so-called *prehistoric* beasts lived together, for man and all the animals, which included the dinosaurs (dragons)[1] were created on the fifth and sixth days (Genesis 1:20–31).

Contrary to popular opinion, which has been fueled by irrational and unfounded evolutionary beliefs, there is no such thing as *prehistoric* men or animals. *Prehistoric* comes from the Latin word for "before," *præ,* and from Ancient Greek ἰστορία (historía). It refers to a time or period that supposedly happened before recorded history. It can also refer to the time period since life appeared on earth, that alleged vast span of time since the beginning of the universe.

The idea and belief that there was a time before recorded history are evolutionary notions and concepts. Generally speaking, there is nothing that has taken place outside of known history. Granted, there are many events that have transpired down throughout history that have not been recorded or chronicled. However, the Scriptures seem to imply, that almost from the beginning, some form of written language was in use. A number of times in the first few chapters of Genesis, Moses, who compiled this book, used written documents already in existence. "*This is the book of the generations of Adam*" (Genesis 5:1). The Hebrew word which we generally translate as book signifies a register, an account or any kind of writing. (See also Genesis 2:4, 6:9, and 10:1.)

Man was contemporary with all creatures (that includes the megafauna) right from the beginning of history. Man did, however, miss the first five days, but God is a reliable witness, and He ordained a written record (the first two chapters in the Book of Genesis) of what took place before man was created. Yes, I realize that according to those who place their faith in evolutionism, people and dinosaurs did not co-exist, but remember this—those are the same people that believe everything made itself, that information materializes out of thin air and then programs itself. They believe that all life came from the materials derived from the crust of the early earth, the rocks, and the material for those rocks just popped into existence once upon a time 13.7 billion years ago. Can anyone really say with a straight face, that those beliefs, notions, and ideas came from rational and coherent thinking?

HISTORICAL ACCURACY OF THE OLD TESTAMENT

Evolutionary historians typically use periodization, which is a way to divide human *prehistory* using the three-age system of the Stone Age, the Bronze Age, and the Iron Age. These categories are further subdivided into other defined eras. Those who use this philosophy to study ancient history are building upon a faulty foundation and premise, an erroneous and fallacious religion: evolutionism. Therefore, most of their assumptions, postulations, conjectures, and dates will always be erroneous. They believe man has risen from non-human brutes with a simple, unsophisticated, or primitive culture; no matter what evidence, artifacts, or relics are found, they will always hammer, batter, adapt, and mold those remnants to fit within pliable pre-existing evolutionary timelines and history. Those ignorant "historians" never take into account the greatest catastrophe that has ever befallen this planet, the flood in Noah's day. Thus, any supposed history that does not give credit to this upheaval will always be very wide of the mark. There is a great store of historical information that confirms the accuracy of the post-flood Table of Nations listed primarily in the tenth and eleventh chapters of the Book of Genesis. Genesis is an historical document of astonishing reliability concerning the post-flood history of this Earth. Ultimately, it is the single most accurate ancient genealogical document known to man.

Professor Robert Dick Wilson (1856–1930) was an American linguist and Presbyterian Bible scholar of vast learning. He spent a lifetime of study to prove the reliability of the Hebrew Bible. He was a formidable opponent of those who held to *higher criticism* views, which hold that Old Testament Scripture is not historically accurate concerning many of its facts, specifics, and dates. In order to thoroughly research and determine the accuracy of the original texts the Hebrew Scriptures had been translated into, Professor Wilson mastered Greek, Aramaic, and Hebrew, in addition to the other 45 languages into which the Scriptures had been translated into up to 600 AD. It has been said of Professor Wilson that he knew more about the Hebrew text then all the destructive critics put together. In 1926 he published *A Scientific Investigation of the Old Testament*, in which he dealt with the text, vocabulary, grammar, history, and religion of the Old Testament. Professor Wilson believed Old Testament Scripture is precise, accurate, and correct regarding the post-flood Table of Nations listed in the Book of Genesis, and so do I. There are a number of other more recent, well researched books[2] that also confirm this.

EARLY POST-FLOOD CIVILIZATIONS AND PUZZLING ANCIENT TECHNOLOGY

When Noah stepped off the ark, he brought with him a bit of antediluvian technology (construction, ship building, and other technologies). The descendants of Noah retained at least for a while some of his knowledge and skills. At the scattering and dispersal of nations during the judgment at the tower of Babel, some of this knowledge was carried with them. This information was used to build the great ancient civilizations that all seemed to sprout simultaneously about 3,000 to 4,000 years ago in Egypt, Assyria, Babylonia, and empires in the Americas such as the Inca, Aztec, and Maya civilizations. These early cultures had advanced metallurgy technologies that enabled them to produce bronze tough as steel and other alloyed metals, and to make enough electricity to manufacture aluminum and to electroplate objects with gold and silver.

These early post-flood civilizations were also able to lift (and move great distances) impressive size blocks of stone (some well in excess of

100 tons), for they had sophisticated lifting machinery at their disposal. You may witness some of their work at ancient cities such as Tiahuanaco in Bolivia and Sacsahuaman, located near the present-day city of Cuzco, Peru. What was so impressive about their stone wall and building construction was their precise cutting and fitting of irregular, oddly shaped blocks of stone, each of which weighed many tons. Many of these earthquake-proof walls still stand after thousands of years. Even today, long after their construction, a sheet of paper cannot be slipped between the blocks.

The rapid advance (from seemingly out of nowhere) of many civilizations in the past is a puzzle to many. In David Hatcher Childress' book, *Technology of the Gods: The Incredible Science of the Ancients*, he discusses seemingly out-of-place, puzzling ancient technology. He presents his case in this introduction:

> ...In this book we will explore the many bits of evidence that lead us to the astounding conclusion that ancient man was virtually as sophisticated as we are today—at least someone, from somewhere, was here using high technology. This technology included everything from electricity to heavy machinery and aircraft.
>
> The topics of ancient flight, ancient atomic wars, ancient electricity and such, will seem odd to many people...yet, as we shall see, there is a lot of evidence pointing to a technologically advanced past. Every culture in the world seems to have legends of ancient flight and a golden civilization before our time.[3]

He continues to explore and ponder these ideas for a few more pages and then he asks this question: "Was the science of Egypt inherited from an earlier culture?"

As he continues with this line of reasoning, he turns to author, scholar, lecturer and guide, John Anthony West, the man who won a News and Documentary Emmy Award for the 1993 television documentary *The Mystery of the Sphinx*. West said this about ancient Egypt:

Egyptian science, medicine, mathematics and astronomy were all of an exponentially higher order of refinement and sophistication than modern scholars will acknowledge.

...every aspect of Egyptian knowledge seems to have been complete at the very beginning. The sciences, artistic and architectural techniques and the hieroglyphic system show virtually no signs of a period of "development"; indeed, many of the achievements of the earliest dynasties were never surpassed, or even equaled later on. This astonishing fact is readily admitted by orthodox Egyptologists, but the magnitude of the mystery it poses is skillfully understated, while its many implications go unmentioned.

How does a civilization spring full-blown into being? Look at a 1905 automobile and compare it to a modern one. There is no mistaking the process of "development," but in Egypt there are no parallels. Everything is right there at the start.

The answer to the mystery is of course obvious, but because it is repellent to the prevailing cast of modern thinking, it is seldom seriously considered. *Egyptian civilization was not a "development," but a legacy.*[4]

Many other researchers, such as Donald E. Chittick,[5] have come to the same conclusions concerning ancient technically sophisticated cultures; no evidence has been discovered that they were developed over a long period of time:

There is no evidence whatever of any technological breakthrough in the methods of quarrying or cutting stone which might account for the onset of pyramid building. All the tools and techniques used by the pyramid builders were in existence well before their time.[6]

The archaeological evidence suggested that rather than developing slowly and painfully, as is normal with human societies, the civilization of Ancient Egypt, like that of the

Olmecs, emerged *all at once and fully formed*. Indeed, the period of transition from primitive to advanced society appears to have been so short that it makes no kind of historical sense. Technological skills that should have taken hundreds or even thousands of years to evolve were brought into use almost overnight—and with no apparent antecedents whatever.[7] (emphasis his)

Numerous are the books that have been written attempting to make sense of the anomalies, the so-called out-of-place artifacts and advanced technologies sometimes found in various locations around the earth. Cryptozoologist Ivan T. Sanderson coined the term *out-of-place artifact* (OOPArt) for archaeological, paleontological, and historical objects that do not seem to fit within conventional historical chronology. These are objects that are supposedly too technically advanced and have a level of sophistication that could not have been attained by the historical cultures with which they are associated. These are artifacts that give the impression that vanished civilizations may have possessed greater technology and had knowledge more advanced than our own. It can also describe articles and items that are found in a time well before humans were supposed to exist.

The Baghdad battery was discovered near Baghdad, Iraq, in 1936. To test the battery myth, MythBuster's twenty-ninth episode (which aired on March 23, 2005) tried it out. After fitting ten handmade terra-cotta jars to act as batteries and linking them together in a series, lemon juice was chosen as the electrolyte; they produced upwards of 4 volts. But that did not answer their question: what did the ancients use these batteries for?

The Antikythera Mechanism, probably designed and constructed by Greek engineers, was recovered in July 1901 from the Antikythera shipwreck located off the Greek island of Antikythera. It is a complex mechanism composed of 30 meshing bronze gears and has been called a mechanical computer. It seems to be designed to calculate astronomical positions and eclipses, as well as the Olympiads, the cycles of the ancient Olympic Games. It has been dated to about 205–100 BC. There is no known predecessor or successor of a machine with such advanced technology. Machines with this kind of complexity would not appear again until the fourteenth century.

There are other such artifacts, namely the Saqqara Bird found in Egypt and comparable objects in Colombia and Ecuador that some allege look similar to modern planes and gliders. But most dismiss those claims, for they contend they are just stylized representations of birds and insects. However, according to a June 9, 2010 article by Doug Aamoth,[8] the History Channel, to prove that the Saqqara Bird was capable of flight, engaged aerodynamics expert Simon Sanderson to build a replica, and when tested in a wind tunnel it flew quite well.

In Dr. Donald E. Chittick's book *The Puzzle of Ancient Man*, he described a pure cast-gold object found in a grave in Colombia, South America, now residing in the Smithsonian Institution in Washington, D.C. It is estimated by the museum to be about 1,000 years old. He wrote:

> Nevertheless, because of its aerodynamic design, particularly the tail section with its high flanged rudder, like the tail on modern planes, the object gives the distinct impression that it is a model of an airplane. Even scientists who have closely examined the object believe that it may be an aircraft.[9]

Another in the long list of challenging artifacts for modern archaeologists is the London Hammer, found by hikers in a creek bed near London, Texas, in 1934. It was embedded inside ancient Ordovician limestone, supposedly 400 million years old. It now resides in Carl Baugh's Creation Evidence Museum, located in Glen Rose, Texas (Baugh purportedly had the hammer tested, but, as some contend, not in a transparent way). The metal of the hammerhead is composed of 96.6 percent iron, 2.6 percent chlorine, and 0.74 percent sulfur, and it does not rust. Supposedly, the head of the hammer is composed of an iron purity that is only achievable with modern-day technology. If you look at pictures of this hammer, you will see on the left side a small shiny v shaped notch which was made by a file. What kind of modern-made iron, with no kind of rust protection, does not rust after 82 years? Some say this hammer is stylistically consistent with American tools that were manufactured in the western region of the United States in the late 1800s. One possible explanation they say is, via common geological processes, the hammer was dropped in a crack or simply left on the ground and soluble chemicals acting in natural fashion hardened the materials around the hammer.

I believe this hammer is post-Flood, three or four thousand years old. There may be a simple method to settle this matter. Analyze the material in metal tools produced in the 1800s in the United States, and see if the composition is similar and has the same qualities as the London hammer. If not, it is probably ancient technology.

There is a lot of controversy concerning this artifact, thusly, it would not seem to help make my case, but there is one very interesting element about it, which is why I presented this information. The hammer was encased in rock! This is not the only example out of many (if the hammer is of recent vintage) of modern artifacts being found sheathed or cemented in rock, coal, or other materials supposedly millions of years old. In another instance, a brass bell with an iron clapper was found in a lump of coal (a nuclear activation analysis revealed the bell contained an unusual mix of metals, different from any known modern alloy production, including copper, zinc, tin, arsenic, iodine, and selenium); there have also been cases of a strand of gold thread embedded in rock, an intricate gold chain found in a coal seam, and a large ceramic spoon, a cut-iron six-penny nail, and a tapered, threaded iron screw all being found in chunks of rock.[10] Is this not a clue that rock does not need millions or even thousands of years to form? Materials that evolutionists insist took long ages to form could in fact have been transformed into rock within just a few years.

Some people, such as Erich von Däniken in his 1968 book, *Chariots of the Gods*, have proposed that intelligent extraterrestrial beings, aka ancient astronauts, have visited Earth in past ages. He probably reasons, as evolutionism is true, why couldn't it be that somewhere in this vast universe, in ages past a race of beings arose, evolved a very high level of skill and sophistication, and visited this planet? It is their technology and knowledge that helped create these puzzling artifacts. In fact, some believe we might be their offspring, for they believe those beings seeded this planet with life many eons ago.

Is the philosophy of evolutionism not wonderful? For it is the solution to the problem. Although the semi-brutish evolving ancestors of modern humans could not make those artifacts, numerous ages before bacteria crept out of an earthly slime pool, a race of super intelligent beings from some far-flung, remote galaxy, beat us to the punch, for they were also able to self-create themselves, develop the technology, and then fly here in time to become ancient man's nannies and nursemaids. In reality, there is no great mystery, for none of those artifacts are out of place. What is found is just as it should be, for Scripture transmits a limited, but accurate, historical record. West was correct: "Egyptian civilization was not a 'development,' but a legacy." It was a vestige of the antediluvian society's technology, which was transported by Noah aboard the ark, and passed down to his descendants.

The problem does not lie with what is found, but with evolutionism's reasonings, analyses, and interpretations, for they are built upon a faulty foundation and premise, an erroneous and fallacious religion, evolutionism. We don't need Erich von Däniken's silly ideas or programs on the History Channel such as *Ancient Aliens* to explain the seemingly out-of-place artifacts and advanced technologies often found in various locations around the earth. Henry M. Morris in his book, *The Genesis Record*, clearly stated why this is so:

> Thus, that first mighty civilization, which once thrived over much of the earth, has been almost forgotten by the world of modern scholarship. In his researches, the ethnologist does encounter stories of a sunken Atlantis or some kind of

mythological Golden Age. In recent years, a number of amazing artifacts have been brought to light by writers advocating the strange notion of ancient interplanetary astronauts.

Actually, these artifacts give, instead, an insight into the remarkable technological skills of early men, some of whom may actually have been antediluvians. Such data are still quite controversial, but at least they do convey the impression that early men were far from the brutish primitives that modern evolutionists have imagined them to be."[11]

CIVILIZATION BEFORE THE FLOOD

For the remainder of this chapter we will be discussing some post-flood aberrations and deviations from the norm of Noah's descendants, and will also step way back into history, to the antediluvian age, the very beginning of time, until the time, *"the world that then was, being overflowed with water perished"* (II Peter 3:6).

What was it like, the first civilization mankind developed, established and formed? What were the people like, their physical stature, their intelligence and capabilities that God gave them to live in the incredible environment, the natural world with its ecosystem of flourishing giants? Might the people also have been correspondingly proportional with all the rest of the natural world? Speaking logically and comparatively, would God have made two midgets to live among a world of colossi? We do have a *witness*, a record, that *"there were giants in the earth in those days"* (Genesis 6:4). Could this verse refer to both, man and the beasts?

Due to evolutionary propaganda and indoctrination, many folks would have a hard time imagining people living in the ancient, pre-flood world. Some might envision a planet like the two-season science fiction television program from the late 1960s, *Land of the Giants*. A sub-orbital transport spaceship from Earth ran into a mysterious storm and ended-up crash-landing on a world with a similar civilization and technological development, they even spoke the universal language of all alien civilizations, English! However, everything was twelve times larger than its counterpart on Earth. It seems all the crew and passengers of that spaceship did each episode was rescue fellow survivors, and avoid being stepped on or captured by the giants. Others might picture Adam and Eve and their descendants scurrying about like ants, darting from place to place in a land of behemoths. Their whole lives consumed with avoiding being eaten by T-rex and his ilk, or being stomped on and squished by massive beasts prowling about, whose sole purpose and mission in life was to seek out and destroy two-legged, furless, big-brained Homo sapiens.

Adam and Eve were probably greater in stature than most folks today. The question is, how much greater? There really is no way of knowing with any certainty. However, it would be a reasonable guess that whatever man's stature is in correlation to living creatures today, would not that same relationship have prevailed in the original world God created?

In the genes of that original perfect substantial pair were all the information needed to produce all the varieties, assortments of traits, physiognomies, and characteristics found within the numerous nations and cultures now abounding upon this earth.

ANCIENT GIANTS

During the time of Moses (about BC 1450), the children of Anak, a race of giants called the Anakim, scared ten of the twelve spies sent to scout out the Promised Land so much that they brought back a malevolent report of the land.

> And they brought up an evil report of the land...And there we saw the giants, the sons of Anak, which come of the giants: and we were in our own sight as grasshoppers, and so we were

in their sight. And all the congregation lifted up their voice, and cried; and the people wept that night. (Numbers 13:33, 14:1)

Israel also encountered Og, King of Bashan, a remnant of the Rephaim, a race of giants. His bedstead was constructed of iron, and was 9 cubits in length and 4 cubits in breadth (over 13 ft long and 6 ft wide (Deut. 3:11). His actual height is not mentioned, but a man that needed something that large to sleep upon would certainly have been someone you bowed to when in their presence and called Sir.

Later, during the time of King David, the children of Israel came across another scattered remnant of giants probably descended from the Rephaim, who took refuge with the Philistines: Goliath of Gath and four of his brothers. These God-defying men who stood against Israel were slain by David, the future king of Israel, and his brothers. A question many ask is how tall was Goliath? The Masoretic text which most English translations follow lists his height as *"six cubits and a span"* (approximately 9' 9", I Samuel 17:4).

This does seem a reasonable figure, considering the sheer weight of armaments which would have required him to be of enormous size and strength. His coat of mail alone weighed 125 lbs, and just the tip of his spear was 15 lbs. This does not even take into account the armor on his legs, sword, javelin, or his helmet.

Goliath had to have been very tall and massive, for he frightened a whole army. *"When Saul and all Israel heard those words of the Philistine, they were dismayed, and greatly afraid ...And all the men of Israel, when they saw the man, fled from him and were sore afraid"* (I Samuel 17:11, 24). Even Saul, king of Israel, was afraid, though he was also a giant of a man and must have stood at least 7 ft tall, for the scripture states: "from his shoulders and upward he was higher than any of the people" (I Samuel 9:2). I find it very hard to believe that a man standing 7 or even 8 ft tall would have terrified every soldier in the army of Israel. But a man trained in combat from his youth, endowed with great strength, weighing in at about 600 lbs, standing 10 ft tall, all decked out in about 200 lbs of armor, with a thunderous voice, mean-tempered, battle-hardened, and itching for a fight; a giant like that could make even the bravest of soldier's hair stand on end, and become a little weak in the knees.

Down throughout history there have been stories, legends, and folklore concerning giants. On the Internet you will find numerous articles and stories about this topic. There are those like Steve Quayle (I certainly do not hold to, nor agree with everything he believes) who have written books on this interesting and somewhat controversial subject. In his book *Genesis 6 Giants—Volume 2*, he deals with legends, myths, and oral traditions about giants that, after you have read his book, may not seem as farfetched as you have been taught or lead to believe. On his website he states:

> Yet the truth remains: Those legends, myths, and oral traditions may not be so farfetched as you have been taught, and even would prefer to think. In fact, they may have at least nuggets of truth in them. That truth points back to a period of time, in the dateless past, when great civilizations flourished with scientific and engineering achievements (which challenge modern intellects) were then the norm.[12]

MODERN-DAY GIANTS

As in the past, there exist in the present people of various shapes and sizes. In the rainforests of Central Africa, there are tribes of pygmies whose adult men average less than 5 ft tall and the median height for women is just 4 ft. There are some Sudanese, Turkana, and Tutsi tribes inhabiting Africa, which probably originated in the Upper Nile area, where heights of 6 ft and above are just average. In the recent past, within the last 100 years or so, there have been both men and woman who stood over 8 ft tall, who could properly be called giants; sometimes the reason for the excess height is due to a glandular condition, but some folks just come from the stock of families who are just big people.

One such couple, the Fischers, who passed away during the 1980s, travelled with circuses such as Ringling Brothers, and were once billed as the largest couple on earth. If you had met them, I am sure you would have agreed, for if they were standing before you, they would have certainly been an impressive, imposing pair. Alfreda's height was listed as 8' 0" and Gottlieb's at 8' 2." They weighed, respectively, 350 and 355 lbs. Another case was that of Vaino Myllyrinne (1909–1963), a soldier in the Finnish army, who at one point in his life (1961–1963) was measured at 8'3", making him officially the world's tallest man. He is also considered the tallest soldier in history. For most people, if they were as tall as Myllyrinne, would not being a soldier be the last kind of occupation or profession they would choose if a long life was on their mind? For when a soldier is in battle, unless he is trying to commit suicide, the last thing he wants to do is standout, be exposed, and be the biggest, tallest, cannot-be-missed target to shoot at.

The giant Edouard Beaupré, at 8'3" was reportedly one of the five or six tallest men in recent recorded history. He was also endowed with great strength. During his performances in the circus freak sideshow of Barnum and Baileys circus, as part of his act he would reportedly lift horses as heavy as 900 lbs to his shoulders. He died in 1904 from tuberculosis at the young age of 23. According to the Guinness Book of Records, the world's current tallest human is Leonid Stadnyk. Born in the Ukraine in 1971, he is 8'5". At 8' 11.1" tall and weighing 440 lbs, the Alton giant, Robert Wadlow, whose height is verified by indisputable evidence, is considered the tallest man in history. He died in 1940 from an infection caused by a blister on his ankle.

There is however, one other man, Johan Aasen, the Norwegian Giant, who might have broken the 9-foot barrier. His official height was listed as 8'9¼", but according to his death certificate from Mendocino State Hospital, at the time of his death in 1938 he was 9'2".

Occasionally in this present world we see giants, but they are an aberrattion and divergence, which is at variance with the "norm" of the rest of this world. Back then, however, everything was brand-new, faultless, and pristine in the world God created, and colossuses were standard and customary models of creatures, man and beast in the antediluvian world and not an anomaly or deviation. So, considering the previous information, would it be unreasonable to assume that those perfect beings, Adam and Eve, might have stood 8 to 10 ft tall?

Robert Wadlow; photo public domain

INTELLECTUAL ABILITIES AND QUALITIES OF THE ANTEDILUVIANS

So, besides their stature, what else was different about Adam and Eve, the father and mother of all mankind? The first pair were created by the hand of God, and thus they were physically and mentally perfect with no genetic defects or abnormalities to pass on to their progeny or heirs (whose imperfections plague much of mankind today). Adam and Eve were created in the likeness and image (Hebrew: *Tselem*) of God. They were a representation of real substance, a shadowing forth of God's intellect, His nature and perfection. They were a masterpiece of divine excellence, astonishing creatures, made in the similitude of God.

Adam and Eve would have had intellects far greater than any fantasy being or alien race dreamed-up, fantasized, or envisioned by evolutionists, for an almighty flawless being, God, is much better equipped to make a perfect human than evolution's "god", better known as Nothing. What would it have been like to have a mind with God-like intellectual abilities and qualities? A mind that was 100 percent functional, endowed with almost limitless capabilities of imagination, creativity, resourcefulness, motivation, cognitive and executive functions?

Adam and Eve would have had no need for the drug taken by Edward Morra, the struggling writer portrayed in the 2011 minor box office success *Limitless*, an American film thriller. Morra was introduced to a *nootropic* drug (smart drug/cognitive enhancer) called NZT-48, which gave him the ability to fully utilize his brain functions, causing him to become a brilliant intellectual virtuoso. Adam and Eve's mental abilities would have surpassed those of Morra, even after he took the drug. The evolutionist's universe is a make-believe, imaginary place of extraterrestrial beings, aliens, and ancient astronauts from far-flung planetary civilizations—but whatever imagined intelligence those fictional beings supposedly possessed, Adam and Eve would have had a mind that surpassed even the Krell, an alien race of super-intelligent beings, from one of my all-time favorite movies, *Forbidden Planet*, a 1956 American science fiction film.

This highly advanced race inhabited the distant planet Altair IV, light-years from Earth, but for some mysterious reason they died out suddenly, destroyed in a single night, nearly 200 centuries before. In the twenty-third century, powered by Hyper-Drive, a faster-than-the-speed-of-light propulsion engine, the starship Bellerophon traveled to that planet with a team of scientists. Dr. Edward Morbius (Walter Pidgeon) and his party discovered the remains of this extinct civilization those prodigies left behind when they landed. However, after a time, one by one, most of their party met the same fate as the Krell. Despite all precautionary measures and safeguards, they were still torn apart limb from limb by what Dr. Morbius described as an "unknown planetary force." It seems that somehow, Morbius and his wife were immune to that menace, but the other terrified survivors attempted to flee on the starship; however, as it lifted off, it was vaporized before they could escape. Twenty years later, another spaceship with a military team aboard, reaches the planet to discover or determine the fate of the earlier expedition.

The Krell, it seems were a benevolent race of beings, although most of the descendants of Adam and Eve were not so—as the Scripture states, *"And God saw that the wickedness of man was great in the earth, and that every imagination of the thoughts of his heart was only evil continually… The earth also was corrupt before God, and the earth was filled with violence"* (Genesis 6:5, 11). The crowning achievement of the Krell civilization was an enormous technological marvel, a machine that could create material and matter in any shape, size, or configuration, by just mere thought, which was powerful enough to supply the wants and needs of a whole population of super-geniuses. This machine was instrumental in the destruction of the Krell; it was the means through which they had unwittingly wiped out their race. As the movie comes to its electrifying conclusion, we learn that Dr. Morbius's subconscious mind had accidently gained access to the Krell's still fully functional machine, and he was now unknowingly wielding its power.

The race of beings called the Krell are science fiction, but those who populated this planet before the flood were real-life flesh and blood, bona fide geniuses, brilliant in mind and superb in body—so magnificent in body that many saw more than nine centuries pass before their breath ebbed away. What would a man such as Thomas Edison, the most prolific inventor in American history, have devised or developed, if he had another 800 years to continue his work? What could just one inventive genius accomplish in his lifetime, if he could continue to learn, grow in knowledge, build upon and expand his understanding of whatever task was set before him, within a lifetime that extended for 900 or more years? What could a genius achieve, whose mental prowess and abilities would not begin to abate until eight or nine centuries had passed? What could a whole race, millions of such long-lived, super intelligent beings develop and create in 1,656 (Creation to Flood) years?

One question that comes up when dealing with the Scriptural genealogy: if individuals lived as long as 900 years, how did so many generations pass between the expulsion from Eden and the Flood? While that may seem to be a good question, and cast doubt on the Scriptural lineage, in reality that question is really not well thought-out.

In my wife's family there are four generations all within the timeframe of 67 years. My mother-in-law is a great-grand-mother, and if she lives another 20 years or so (she is in relative good health, so, God willing, it is possible), she could very well be a great-great-grand-mother, with five generations all living within the span of one person's lifetime of less than a century.

Adam lived for 950 years. Methuselah was born while Adam was still alive, and attained the oldest recorded age listed in Scripture, 969 years. Here is a case where two men's lifespans together cover more than 1,600 years. As recorded in Scripture, there is no problem to have numerous generations exist within the lifespan of both men.

Consider a few things that ephemeral, contemporary man, a fairly tarnished, degraded and inferior copy of a copy of a copy of God's perfect original handiwork, has accomplished during just the last 200 years or so. In 1804, the first steam locomotive, made by British inventor and mining engineer Richard Trevithick, went into operation along the tramway of the Penydarren Ironworks in Merthyr Tydfil, Wales. Philo Taylor Farnsworth designed the first electronic television, which was successfully demonstrated in San Francisco on September 7, 1927. By 1939 RCA televised the opening of the New York World's Fair and broadcast a speech by President Franklin Delano Roosevelt, the first president to appear on television. Later that same year RCA began selling television sets with 5" by 12" picture tubes. The poor quality of the television image at first hindered its marketing, but the quality of the picture continued to improve, and by 1947 in the United States, full-scale commercial television broadcasting began, just a scant six years before I was born.

On August 27, 1939, Erich Warsitz piloted the first practical jet aircraft, the Heinkel He 178. The development of jet-powered flight advanced so quickly, that one year after I was born, the first commercial jet passenger airliner, the Boeing 367-80, took flight on July 15, 1954. It was a prototype that developed into the 707 airliner, which would revolutionize commercial air travel. Four years after I was born, the first artificial satellite, Sputnik, was launched into low Earth orbit by the Soviet Union on October 4, 1957. This surprise success triggered the Space Race, and in the ensuing years ushered in new political, technological, military and scientific developments. For those my age, most importantly, in 1957 General Foods Corporation introduced Tang, a fruit-flavored drink mix. It became a very popular drink after NASA used it on John Glenn's Mercury flight and subsequent Gemini missions.

I grew up before cable, satellite, and color TV, remote control, cell phones, personal computers, laptops, DVDs, Internet, 3D printers and so much more, things my children find just a normal part of their lives. Consider what could be accomplished with a world populated with innovative geniuses such as Nikola Tesla (1856–1943), the inventor (holder of over 300 patents), electrical and mechanical engineer, physicist, futurist, and polyglot with an eidetic and photographic memory.

What kind of apparatus, tools, machinery, and technology did those brilliant antediluvians invent? With all of their cunning, savvy and astuteness, what did their material culture consist of, look like? What kind of devices, widgets and gizmos came from the minds of those akin in intellect, mentality, and spirit to Tesla?

Each year I used to get the Encyclopedia Britannica *Book of the Year*. Each volume contained about a thousand pages of the most newsworthy events from around the world the previous year. Of course, even that one volume could hardly report more than just a few highlights of supposed historical note that some journalists considered important. Imagine how fascinating it would be to find a newspaper from a large city of the antediluvian civilization, or a *Book of the Year*. To read the actual accounts of people, things, and events that occurred day to day a thousand years after Creation. Of course, it would probably take a philologist like Dr. Morbius to help translate it for us. (Something else that we should note is that Dr. Morbius and his ilk would have been of little use in Noah's lifetime, for the entire earth *"was of one language and of one speech"* (Genesis 11:1). However, I have a very deep suspicion that if an evolutionist found an artifact like a newspaper, and it could be confirmed it was a relic dating from the antediluvian civilization, no one other than his or her evolutionary colleagues would ever hear about it, for to protect their religion, it would never see the light of day. It would vanish quicker than a large roast *lechon* (pig) at a Filipino barbeque or birthday celebration.

Very little is revealed or disclosed in Scripture concerning the antediluvian civilization (which lasted for nearly 17 centuries), which developed in the world and was completely annihilated by the flood. However, there are a few tantalizing hints and clues revealed in Scripture. Two of the oldest professions, agriculture and animal domestication, were quickly undertaken by the early descendants of Adam. At the dawn of history, in the first few chapters of Genesis, metallurgy, musical instruments, cities, urbanization, written communication, expertise in craftsmanship in wood and other materials, and mathematics are all mentioned or alluded to. Along with these advanced endeavors and undertakings, trade, commerce, engineering, industry, and manufacturing would also have been commonplace. However, if this is true, why do we not find more human artifacts pre-dating the flood? If there were great civilizations, if the earth was teeming with millions of human inhabitants, where is the evidence, their remains, their bones, and effects? Where are their edifices and monuments they surely must have constructed, the ruins of their great cities, with all of its structures and buildings? Sometimes evolutionists will use these kinds of arguments as proof that the Book of Genesis and its account of the flood is a myth, just a made-up story, legend or fable.

I have thought about and pondered some of these kinds of questions, interrogations, and points for years. At times they were seemingly unsolvable problems for me and other students of Scripture. These questions do not falsify or disprove anything, but, it would still be nice to have a satisfying, adequate, and reasonable answer. The next few chapters explore some of these questions and provide a few possible solutions and answers to some of these perplexing issues and matters.

NOTES ON CHAPTER 12

1. You will not find the word "dinosaur" in Scripture, for most English translations such as the King James Version (1611) were made long before the word was coined. Dragon is the term used in Scripture for those ancient reptilian beasts. It comes from the Hebrew word *tannin*, which means a marine or land monster, a vast fish or serpent. It has been rendered various ways in Scripture, such as serpent, dragon, whale, sea-monster, and jackal. You will find the Hebrew word tannin used right in the Book of Genesis the first chapter, verse 21 where it reads "And God created great whales [*tannin*] . . ."
The name dinosaur was coined by Dr. Richard Owen, who first used the word in his *Report on British Fossil Reptiles* presented in 1841 to the British Association for the Advancement of Science. The word dinosaur is a combination of two Greek words: *deinos* (fearfully great) and *sauros* (a lizard). Thus the name literally means "fearfully great lizard."

2. There are a number of good books dealing with this subject, including *Ancient Post-Flood History: Historical Documents that Point to Biblical Creation* by Dr. Ken Johnson (2010), and (in my option) one of the best, *After the Flood: The*

Early Post-Flood History of Europe Traced Back to Noah by Bill Cooper (1995).
3. David Hatcher Childress, *Technology of the Gods: The Incredible Science of the Ancients* (Kempton, IL: Adventures Unlimited Press, 1999) p. 9.
4. Ibid., p. 12.
5. Donald Chittick, *The Puzzle of Ancient Man* (Newberg, OR: Creation Compass, 1998), pp. 78–79.
6. Kurt Mendelssohn, "A Scientist Looks at the Pyramids," *American Scientist*, March-April 1971, p. 210.
7. Graham Hancock, *Fingerprints of the Gods* (New York, NY: Crown Trade Paperbacks, 1995), pp. 135–136.
8. Cryptids: The Saqqara Bird | TIME.com, http:// techland. time. com/2010/06/09/cryptids-the-saqqara-bird/
9. Chittick, 1998, p. 109.
10. "Impossible Stuff in Coal and Rock," 6000years.org web site, n.d.,www.6000years.org/frame.php?page=stuff_in_coal
11. Henry Morris, *The Genesis Record* (San Diego, CA: Creation-Life Publishers, 1976), p. 42.
12. Steve Quayle, "Genesis 6 Giants," Steve Quayle web site, http://www.stevequayle.com/index.php?s=621, para. 21.

> It must be significant that nearly all the evolutionary stories I learned as a student, from Trueman's Ostrea/Gryphaea to Carruthers' Zaphrentis delanouei, have now been 'debunked'. Similarly, my own experinece [sic] of more than twenty years looking for evolutionary lineages among the Mesozoic Brachiopoda has proved them equally elusive.*

CHAPTER 13

THE GROUND BENEATH US

As you read the next few pages, you will find that my retort is not proof, undeniable scientific evidence, or irrefutable as to the accuracy of my riposte. How could it be, for these events we are discussing are history, and therefore are not repeatable, reproducible, nor testable? It cannot be proved or verified that they happened by a certain prescribed manner by any scientific demonstration. This is the exact same problem evolutionists face when dealing with their religion. I am seeking a reasonable answer, not the definitive answer. However, my response will be vastly superior to theirs, for what you will be reading is a sensible (scientific), logical, and Scriptural response to the mockery of ill-informed, ignorant, and oblivious individuals.

Now, we may be able to prove that it is scientifically possible for things to happen and occur as we suggest, but that does not verify that that is *the* way the past events happened or transpired. In our studies and research, even if we unknowingly happened upon the actual reason, cause, and way events really took place, we will never be able to

---
*Dr Derek V. Ager (Department of Geology & Oceanography, University College, Swansea, UK), 'The nature of the fossil record'. *Proceedings of the Geologists' Association*, vol. 87(2), 1976, p.132.

verify beyond a shadow of a doubt that that is the correct answer to our inquiry.

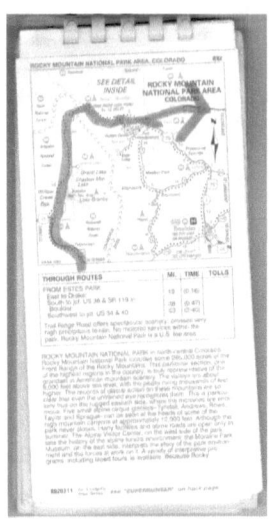

In this chapter we seek the answers to queries such as, why has so little evidence been found from the antediluvian civilization? Is there really enough water to cover mountains such as Mt. Everest (hint: *all* the world's mountains have not always been as tall as they are now)? If the whole world was flooded, where did the water go after the Flood? To answer those and other analogous questions, we will deal at length with one particular issue, the terrain right under our feet and the flood that formed it.

After my fiancée Maribeth and I got married in 2004, our honeymoon was a nine-day, 4,400-mile round-trip journey out west to see the Petrified Forest, Bryce Canyon, Grand Canyon, and other sites my wife and I had only gazed upon in pictures. After we stopped by AAA and got our free Trip-Tik, a nice spiral-bound set of maps in which the clerk highlighted the route we would use on our journey, we left Milwaukee. Most of what we saw during the first two days, the forests, hills, bluffs, and other landscapes are not unique to Wisconsin, but are common throughout most mid-western states. It was not until we approached the great state of Texas (about 1,400 miles southwest of Milwaukee) that we saw something different.

There had been a gradual transition of the terrain, so that by the time we drove west from Amarillo, Texas, it had become, for as far as you could see in every direction, flat as a pancake; there was not a hill, valley, or even a grove of trees to break the monotony of the view. It is so flat that, as they say down there, you can see your dog run away for three days.

On the third morning of our trip the sun was peeking up over the horizon when I looked out the hotel window. I enthusiastically woke my bride and said (what became a common refrain during the rest of our trip), "Honey, look at this, it's something you have never seen before." The sun was melting the frost on the windshield of our car. Being from the Philippines, she had never seen frost or touched snow in her life. About a week later, in the mountains of Colorado, she got to experience what it was like to make a snowball—and to get hit with one.

Maribeth's first snowball in Colorado September 2004

We were very happy to be leaving windswept Amarillo that morning, for that was the day, after driving so many miles, we would finally get to see the first site marked on our map, the Petrified Forest. We were not sorry to leave behind the previous nights' accommodations, our stay in a half-star hotel, which scant amenities included a very used bar of soap, leftover remnants from a small brand-less bottle of shampoo, and free continental breakfast (a tiny can of juice and tasteless dry doughnut). After procuring a second breakfast, we headed west on I-40. Half an hour later were excited to see, gradually emerging from the dreary countryside, that vast treeless plane, the familiar scenery one is accustomed to from watching old Western movies such as the classic 1956 American film *The Searchers*, starring John Wayne and Jeffrey Hunter. From that point on, we began to see scenery and landscapes we had never viewed before.

SEDIMENTARY ROCK

Consider, before we proceed, the ground that lies under our cities, towns, and paved roadways we travel upon. *Sedimentary rock* is formed at or near the Earth's surface from pre-existing rock that had eroded (physically disintegrated). The eroded material is then deposited as sediment by water, and over time the sediment is consolidated by pressure or some kind of cementing agent. Considering the size of the Earth, the sedimentary rock layers comprise a relatively small portion of the Earth's mass. The blanket of sedimentary rock makes up about 5 percent of the volume of the terrestrial crust and about 1 percent of the Earth's total volume. The crust of the earth is dominated by igneous and metamorphic rocks, which lie directly under the layers of sedimentary rock.

Bryce Canyon

The Petrified Forest

The Petrified Forest

The *Encyclopedia Britannica* stated this about the earth's crust:

> What is called *bedrock* can be made of three types of rocks: sedimentary, igneous, and metamorphic rock. While igneous and metamorphic rocks constitute the bulk of the Earth's crust, sedimentary rock blankets 80–90 percent of the surface area of the Earth, rather than igneous or metamorphic varieties.
>
> The sediment-sedimentary rock shell forms only a thin superficial layer. The mean shell thickness in continental areas is 1.8 kilometres; the sediment shell in the ocean basins is roughly 0.3 kilometre. Rearranging this shell as a globally encircling layer (and depending on the raw estimates incorporated into the model), the shell thickness would be roughly 1–3 kilometres.[1]

For our purposes and considerations, in this chapter we will not reckon sedimentary rock as bedrock, for most of it is not part of the original perfect creation. It is composed mainly of previously existing igneous and metamorphic rock, as well as other kinds of minerals and elements which were eroded during the flood and deposited as layers of sediments.

In most places our earth is covered by soil, sand, clay, and various types of vegetation which conceal the rock beneath. On our honeymoon trek, the further west my wife and I drove, the more sparse became the soil and vegetation and more rock was exposed, until in some places, for as far as we could see in any direction, the scenery was composed entirely of sedimentary rock, thinly peppered with small piles of sand, dirt, or rocks. The bare rock was not an unwelcome sight, for, as by some unseen artistic hand, the brown- and russet-toned layers of rock were beautifully crafted and arranged. Sprinkled here and there throughout this vast region was an occasional bush or hardy succulent type of plant. It was a pleasant panorama of picturesque hills, buttes, and colorful valleys that greeted us each mile along the way to our destination. What many people may not realize is that the sedimentary rock which we drove over on our honeymoon trip covers all the earth's continents: It does not stop at the shore of each land-mass, but continues on under the seas and covers the majority of Earth's surface, like the layers of an onion. Keep this in mind as we continue: *all* sedimentary rock covers up the original topography and terrain of the world God created. Most of the antediluvian world is now covered by thousands of feet of sedimentary rock. This means that all those sediments exposed at the Grand Canyon, which are over a mile deep in places, and much deeper in others, are now resting on land, rock, or ground that was once the surface of the earth. Sedimentary rock hides the ground Adam and Eve walked upon, the land the dinosaurs wandered over and the shipyard where the ark was constructed. It now covers the land where soil, forests, rivers, animals, gardens, homes, farms and cities built by the antediluvians would have formerly resided; it also covers the original seabeds of all oceans and lakes which were once scattered over the entire surface of the earth.

All of the sedimentary layers and surface features now present on this earth were probably not formed by the flood of Noah's day. There was probably a great deal of geological work accomplished on the third day, as the land rose up out of the waters that covered the entire earth. This process would have caused a great amount of erosion and redisposition of surface materials as the waters flowed into their newly created basins. Of course, this sedimentary rock would have been totally free of organic matter and fossils, for organic matter comes from deceased organisms, and death did not occur until sin had entered into this world.

Evolutionism teaches that each successive layer of sedimentary rock denotes an age, era, or epoch in earth's history that lasted for millions of years. Although only a portion of earth's history is found within those layers, what is there embodies numerous eons, millions of years. Continents are constantly being worn down by the effects of rain. Those sediments are washed into river systems such as the Amazon, Mississippi, and Congo. At the mouths of these rivers where they encounter the ocean, a continuous rain of sediments falls upon the ocean floor. But how would this process of erosion and weathering form or create the substantial layers of sedimentary rock that contain enormous fossil graveyards?

Everywhere the rock is exposed, we find many layers of sediment thousands of feet thick, neatly stacked like pancakes stretched out in every direction for hundreds of thousands of square miles, generally with no evidence of erosion between most layers. Where did those deposits come from that settled down upon each succeeding layer of sediments? They had to have eroded from somewhere—certainly not from river deltas, for they do not create these kinds of sedimentary layers. Remember, most layers cover an enormous patch of land, and some layers are greater in area than most continents.

FORMATION OF STRATA AND FOSSILS

Nowhere is this process of layer formation taking place today on a small local scale, much less is there world-wide deposition. Nowhere on earth do thousands of square miles of ground stay undisturbed by the forces of nature for tens of thousands or millions of years while sediments slowly build up over time to cover dead creatures. Nowhere, when an animal dies, does its carcass just lie upon the ground undisturbed, waiting for thousands of years in which sediments slowly begins to cover it. This never happens, except in the fantasy, dream-world of evolutionists such as Mark Norell. He is the Chair of the Paleontology Department at the American Museum of Natural History. In an interview, reporter Dan Nosowitz asks our illustrious evolutionist: "Where, pray tell, is the best place for aspiring fossils to die?" Norell answers:

> "You can't really predict what's going to happen in the future," says Norell, "but an ideal kind of place would be someplace out on the Great Plains." It's got everything: it's tectonically stable, well-drained, with few major rivers running through it. It's likely to stay the way it is for quite a long time. "Those sorts of habitats, we know from excavating animals all over the world, are most likely to preserve fossils," says Norell.

At the end of this story, Nosowitz parroted Norell when he stated:

> If you want to become a fossil, and achieve textbook fame in tens of thousands of years and get studied as an educational example of bad ecological behavior, die intact, and be buried somewhere in the Great Plains. And make sure you have all your teeth.[2]

This nonsense is just another case that what is observed in the natural world does not fit with evolutionary beliefs. Those kinds of habitats do not preserve any fossils, for the fossils we find are usually in water-laid deposits.

Naturalist Ernest Thompson Seton estimated that when Columbus landed, about 60 million bison roamed the North American continent. Their range extended from the Great Slave Lake in Canada's far north to Mexico in the south, almost to the Atlantic Ocean and to eastern Oregon. Their vast herds were part of the largest convergence of wild animals that have ever existed on this earth. Over the centuries, hundreds of millions of those beasts lived and died. Yet today, if you were to walk the Great Plains, the very spots where innumerable herds of bison grazed, walked over and spent their entire lives, which Norell believes is an ideal place to preserve fossils, you will find no bison fossils from the herds which recently roamed the land during the last 4,000 years. However, researchers have collected bison teeth, from North and South Dakota, Montana, Wyoming, Nebraska, Kansas and Oklahoma. They also have retrieved bones, horns, and hides of these beasts used by the very parsimonious Indian tribes that once inhabited the Great Plains region.

Located in northern Tanzania and extending to south-western Kenya, African's "Endless Plains," the Serengeti ecosystem covers approximately 12,000 square miles. It accommodates the largest terrestrial mammal migration in the world. Each year about 250,000 zebra, 1.5 million wildebeest, and about 500,000 Thompson's gazelles, eland, and impala make a great circular migration that covers about 1,000 miles. Each year, along with numerous other animals, about 250,000 wildebeest die during their migration due to hunger, exhaustion, thirst, or predation. When a wildebeest dies (if not eaten by predators), its soft tissue will decompose in about two to ten weeks. Sometimes along rivers and other places large collections of bones may be found and take as long as seven years to decompose, while yielding a good bounty for the environment, a long-term source of phosphorus and other nutrients. Throughout the ages that this has been occurring in a supposedly ideal location to preserve fossils, how many do you reckon the immense numbers of modern dead animals have produced? The numerical digit zero appears to be the correctly assigned value. The fossils they find from the rusingoryx and other species have no connection to the present-day herds.

I have personally observed how quickly nature disposes of dead animals. On my parents' 40 acres of wooded land near Princeton, Wisconsin, late one spring I noticed that a medium-size white-tailed deer had met its fate. It lay undisturbed in the underbrush all summer; by fall only the skull, teeth and part of the back-bone with a few ribs were left. The skull lasted until the next summer, but by the autumn of that year, not so much as a tooth was left. It was totally gone in less than sixteen months.

The same rapid decay process occurs in the relatively shallow coastal ocean regions (less than 3,000 ft or so). In those regions the water is generally warmer, and the carcasses may float due to decompositional gases, thus hastening the decomposition. Next time you happen to watch a nature program dealing with the ocean, notice the seabed, you will see no dead fish or any other kind of animal carcass just lying around, waiting to be covered up with sediments so it can fossilize.

Of course, conditions surrounding the decay process do vary greatly in the oceans depending upon the depth. At about 3,300 ft begins the bathyal or abyssal zone. When a large creature, such as a whale sinks to the bottom in this zone, what is called a *whale fall* will occur. Its large carcass, after a period of time, creates a complex localized ecosystem; a certain boon to the scavengers, for complete decomposition may take decades. Whale falls are able to occur in the deep open ocean due to cold temperatures and high hydrostatic pressures which increase gas solubility, allowing the carcass to remain intact and sink to the bottom. This process, though it takes years, does not produce fossils.

The same rapid decay also occurs in lakes. However, as in the deep ocean basins, at times the colder water temperatures close to the bottom of a lake can preserve logs. A booming log salvage business has sprung-up within the last thirty years. When the logging industry was in full-swing during the 1800s, logs were often stored in floating masses in rivers and lakes, and many sank. Instead of attempting to recover the logs, it was often cheaper and much easier just to cut down more trees. Thus, on the bottom of lakes and rivers around the world lies sunken treasure, millions of ancient logs (many hundreds of years old), just waiting to be recovered (After drying and cutting, the prices are often 10 times as much as new wood, and Birds-eye maple, with its iridescent grain can go for $80 or more a foot after its cut.). These logs are not fossilized or putrefied, just preserved.

Deep cold and still, oxygen and light deprived water can also preserve skeletons of people and animals for thousands of years. In 2007 in the sprawling cave system in Mexico's Yucatan Peninsula, three Mexican cave divers were exploring an underwater cavern.[3] In an underwater chamber 164 ft deep were the remains of a young woman, and 26 large animal species, such as cave bears, giant ground sloths, and sabre-tooth tigers. However, these processes do not produce fossilized remains.

One reviewer attempted to prove wrong my contentions concerning the remains of bison and other modern (last few millennia) plains dwelling animals. In response, I turned to the research she thought refuted my arguments. However, as I read through this suggested information, I found, as is all research done by evolutionists, biased, predisposed, and skewed findings, discoveries, and conclusions. How can it be otherwise, when their mind is already closed, and unprepared to admit other possible findings to the evidence at hand (for a Flood and young Earth)? That bias is the faith that evolution happened, and a timeline that continues unabated back through the course of millions and billions of years. Here are a few samples from some of the referenced articles:

> Bison roamed North America from Alaska to Mexico going back 200,000 years.[4]

> Approximately 25,000 years ago the genus Bison passed from Asia, over the Bering Strait land bridge, to North America. Fossil bison from this era were twice the size of modern day bison, weighing around 5,000 pounds with 6-foot horn spans!

Bison adapted well to the environment of the North American Great Plains, and flourished in huge numbers (an estimated 60 million animals were present in the 1700s).[5]

New research by Beth Shapiro of the UC Santa Cruz Genomics Institute and University of Alberta Duane Froese has identified North America's oldest bison fossils and helped construct a bison genealogy establishing that a common maternal ancestor arrived between 130,000 and 195,000 years ago, during a previous ice age... "Bison arrived in North America and quickly came to dominate a grazing ecosystem that was previously reigned over by horses and mammoths for one million years," said Shapiro.[6]

Not a lot is known about the rusingoryx species, but the study of several fossil skulls discovered on Kenya's Rusinga Island have shed some light. The fossils date from between 55,000 to 75,000 years ago, when the hoofed grass-eaters would move around the African plains in large herds.[7]

It makes no difference wherever this kind of research is done, North America, Asia, or Africa, evolutionists always miss the very large discontinuity in the history of Earth, the Flood, which occurred four or five thousand years ago and completely changed the whole course of world history. Any fossils found (such as the enormous fossil bison) are not from modern herds of animals. The evolutionists' manner and methodology of research is illustrated in this short story (think of the detectives as Ph.D.-endowed professors of evolutionism).

CASE OF THE VANISHING PERPETRATOR

James and Kevin were partners with long, distinguished careers as detectives, with credit for solving some of the most difficult cases in the history of the city. Their area of patrol and jurisdiction encompassed some of the most crime-ridden neighborhoods in the City of Chicago, and thus many times had come across horrific crime scenes. As they headed home after their shift, an urgent call beckoned their services. Something terrible had occurred at the newly dedicated Richard Daley Zoo Research Facility. They cancelled their trip home and headed right over. They met the first responders just as they had stepped out the front door. Both were visibly shaken as they described to the detectives what they had witnessed when they went into the building. As the detectives turned to enter the building, one of the officers gently put his arms around their shoulders, to prepare them for what they were about to witness, said softly, "You have never seen anything like this before."

Fifteen bodies lay sprawled out, bloodied and crushed. A few were lying at the foot of a wall, crumpled like rag dolls tossed against a hard surface. Large round bloody footprints made winding trails throughout the facility. From the tracts and force of violence done to the bodies, it was clear a human was not the killer, but what kind of animal could do such a thing? The first responders had found one survivor, a security officer, who with his dying breath managed to tell them he was able to empty a whole clip of armour piercing rounds from his Glock 19 9mm into the face and eyes of the beast before it got him. The detectives wondered why the other people didn't try to escape, for the exits were close, clearly marked, and readily accessible. At least a few should have been able to escape the fate of the others. After a bit of investigating, they found the answer: That night during a storm, lightning had shorted out the security system and fried the door locks, rendering them inoperable. They also found something strange—a very large, gray-colored, bloody, odd-looking, leather-covered mound situated near the main research section. There was what looked like a thick rope with a frayed end sticking out one end, four, slightly bent in the middle, tree-trunk-size poles with bloody flat round bottoms that were affixed to the underside. On an adjoining smaller mound, two large, somewhat round flaps of leather were firmly attached on either side of the mound. A long, thick, flexible and wrinkled hose-like thingamajig was protruding from the other smaller mound between two long white curved objects with pointed ends. Whatever that heap was, for the remainder of their investigation the detectives never gave it a second thought. In fact, numerous times they climbed its bulk to get a better perspective as they surveyed the crime scene, and took pictures when perched on its topside. And when they took an occasional break for a bite to eat, they would climb to the top with a chair and sit-down. Even after a few weeks, when the stench of rotting flesh began to emanate from the mound and swarms of flies began to buzz around, they were never interested in any investigation of that large leather wrapped mass.

After spending thousands of hours pursuing every possible lead, checking out any and all tips, and conducting every kind of test and analysis, for only the third time out of the hundreds of cases they had investigated, they had to admit failure. They decided to head back to their offices and write-up the report. Their wives would be so relieved to have their husbands back, for both detectives were so adsorbed in this case, there were days they never left the crime scene, and they ate and slept at the Research Facility.

Back at the office they wrote their final remarks and comments of their investigation, and summarized it this way: Only a very large animal could have caused the carnage and damage done to the interior of the building. Only two living animals have feet which match the trail of footprints found, the hippopotamus and rhinoceros. However, no carcass of any kind of beast was ever found. The animal/s responsible did not escape or exit the building, for the reinforced steel doors showed no damage and were still locked when the police first arrived. There were no holes in the walls large enough for even a cat to get through, much less a four- or five-ton beast. But the strangest thing was the animals' trail of bloody footprints: They stopped abruptly at the mound, and its body was not found, it just vanished, how or why is not known. The case still open.

Long after the detectives had retired, at times they would look into the files and try to puzzle out this mystery. Numerous other researchers took-up the cause in an attempt to solve this baffling enigma. It remains unexplained to this very day.

It is not a mystery. The evidence is clear: They are blinded by an idea, evolutionism. When evolutionists do research out in the "field" they are literally standing upon a graveyard that modern day processes can't reproduce. They walk around on the elephant (water deposited rock strata), climb up and down on its carcass, and handle its remains (fossils), but remain oblivious to the elephant, even though its rotting stench is overpowering. When reading research studies conducted by evolutionists, one is always guaranteed to come across poisoned conclusions.

When an animal or fish dies it never fossilizes, unless something very unusual happens—sudden and rapid burial by large amounts of sediments carried by flowing water, which contains dissolved cementing agents. This kind of very special condition or circumstance required for world-wide fossilization only happened one time in recorded history, about 4,500 years ago, during the lifetime of Noah.

The following accounts are just a few well-known occurrences found by geologists of the innumerable times fossils like this are found. These exceptional conditions buried a 6-foot-long marine reptile, an ichthyosaur, found fossilized at the moment of giving birth. One moment this creature was giving birth; seconds later, without a chance to escape, she and her baby were buried in a catastrophic avalanche of mud. These circumstances buried alive so quickly, fish, which are now fossilized, in the process of eating their last meal. In Fossil Lake, Wyoming, Princeton University scientists found a fossilized perch swallowing its meal, a herring. Their death by sudden burial happened so rapidly that the perch did not even have time to finish its snack; its prey's tail was still sticking out of its mouth. We find fossils of animals buried so quickly that not only do we find food in their stomachs, even the soft parts; the delicate flesh imprints of their bodies are visible.

In Texas huge masses of fossilized clams have been found packed so tightly together it is as if someone was getting them ready to be packaged and shipped somewhere. What is unusual about these clams is that their shells are closed! When a clam dies its shell opens, so, to prevent those shells from opening, they would have had to be quickly buried under the pressure of many feet of sediment. But evolutionists affirm just the opposite. In reference to the geology of the Grand Canyon region, deposition and accumulation of sediments, which formed sedimentary rock (the layers range in thickness from a few inches to over 1,500 ft), supposedly would have had to happen over millions of years, for as Dr. Ken Bevis, a Professor of Geology at Hanover College in Hanover, IN. states: "The sedimentary geology of the Grand Canyon region is extremely diverse and spans more than a billion years of earth's history."[8] While they do not believe deposition was continuous, still, each layer would have needed millions of years to form while those creatures were slowly entombed as the sediments accumulated over their dead bodies.

Fossils for the most part are composed of once-living creatures, plants, and animals, and sedimentary rock is the final resting place, the extensive world-wide tomb where the majority of fossils are found. The sheer immensity and number of these fossil graveyards is incredible, and yet it never happens today! With the sole exception of Noah's flood, throughout all of man's history, this process has never been observed, witnessed, nor recorded by man.

WHAT LIES BURIED DEEP BENEATH OUR FEET?

How much of earth's immense ossuary have the archeologists, paleontologists, and anthropologists excavated, quarried, and exhumed? Not much, for ultimately in due course, all they have done is little more than slightly scratch, scrape, and probe a few square miles of the surface of this planet. How many of those first few layers of deposited sediments, now transformed into stone, resting upon the bedrock, have been scrutinized, inspected, and studied? Considering that sedimentary rock is a world-wide phenomenon and occurrence, and excavating thousands of feet of material to uncover the surface of the bedrock beneath would be a mammoth task, virtually none of it has been surveyed, much less seen!

Seventy-one percent of the earth is now underwater. In the Pacific Ocean, there are places like the Mariana, Tonga, Kuril-Kamchatka, and Philippine Trenches, which dip more than 34,500 ft below sea level. Sedimentary rock lies compressed under the sea floor of those trenches. But what those layers rest upon (which might have been the surface of the land that Adam and his next of kin walked upon) has never been explored, nor is it likely to be. The remaining 29 percent of earth's surface is above the water, but much of that is also inaccessible to exploration. What lies under the thousands of feet of sedimentary rock which is tucked under the ice and snow in the arctic region, and under the 6,400,000 cubic miles of ice which covers much of the continent of Antarctica? How much of the land that lies beneath the thousands of feet of sedimentary rock covered by 677,855 square miles of ice and glaciers on the world's largest island (or smallest continent), Greenland, have been explored? None of it would be a good conservative estimate. Most of the top 4,400 ft of Mount Everest is composed of sedimentary rock, where the remains of sea life such as corals can be found. What lies beneath these layers that were once under water and now lie upon a layer of granite, which overlays layers of shale, gneiss, and igneous rock? It is a reasonable guess that 99.999 percent of what is under the first mile or so of sedimentary rock, of the current surface of the earth, which would probably be where the artifacts of the antediluvians would be found, has never been explored.

The evolutionists are clueless as to what lies deep beneath the earth that their universities are built upon, where they profess their allegiance to evolutionism. Other than a few mines, deep cracks (such as the Grand Canyon), and other fractures, crevasses, or subsurface features, which give us just a peek, a glimpse into what is underfoot, almost nothing is known what of lies beneath the very ground our cities and homes happen to be built upon. Is it not possible that the unexplored terrain buried thousands of feet right beneath your house, or mine, could be the very ground where Noah constructed the Ark, or where the Garden of Eden could have been planted? Why not?

The flood swept away all antediluvian cities and towns and eradicated all traces of their civilization. Considering that such a small portion of the surface of this earth has been explored, is it not possible that just maybe, somewhere in the vast expanse of unexplored land buried beneath thousands of feet of sedimentary rock, some relics and vestiges of the antediluvian civilization are waiting to be unearthed and brought to light?

CAPTAIN NOAH DISEMBARKS

Some writers say Noah was a skilled navigator, but Captain Noah needed no navigation skills, for there was no itinerary, route, or planned destination at the end of the voyage. The Ark had no sails to help speed it along, or a movable rudder to help guide or steer the ship. It was nothing more than a floating cargo barge, totally at the mercy of the wind, waves, and currents, but under the watchful eye of the Creator. After floating for months, drifting with the wind and currents, where the Ark finally came to rest would have been hundreds, probably thousands of miles from where Noah's dry-dock had been. If by some miracle the Ark had come to rest at its departure point, Noah's home, his city, or village would now be buried under thousands of feet of sediment.

When the Ark finally did make landfall, if, by using the stars, Noah was able to determine his location (assuming the continents had not shifted a great deal), why would he have a need to do so? As long as his ship did not dock in the middle of Antarctica or other inhospitable land, would it really have made any difference where he ended up? Of course God had a reason and purpose for depositing the Ark where it landed, but Noah had no choice as to its resting place after his voyage.

Since the time Noah disembarked from the Ark, many have wondered and have asked the question, "Where was the land of Eden where the Garden of Eden was planted?" The disappointing answer—no one has a clue! When Noah walked the earth, all four original rivers—the Pison, the Gihon, the Hiddekel (Tigris), and the Euphrates—that flowed out from the Eden river to water the Garden of Eden, were probably still in existence. In the Middle East there are rivers that people have named after two of the rivers that flowed through the Garden of Eden, the Euphrates and the Tigris. Same names, but no relationship with those famous streaming quadruplets that flowed through the land of Eden. The present-day Euphrates-Tigris river system has its source in the Taurus Mountains of eastern Turkey. The rivers flow in a south-easterly direction through/by the countries of Syria and Iraq and converge shortly before flowing into the Persian Gulf. The Taurus Mountains, which are the source for those two rivers, did not exist before the Flood. Therefore, those current rivers are not part of Eden's river system.

Most scholars have an opinion, and many Bible commentaries state that the Garden of Eden was located somewhere in the Middle East, situated somewhere close to where the Tigris and Euphrates Rivers flow today. There has been much speculation, time spent in research, and books and articles written on the possible location of the Garden of Eden. However, many do not realize the intensity nor the power of the Flood to rearrange the surface of the earth, to alter and completely change its geology, topography, and appearance. Often a small local flood experienced by a community or town can wash away all landmarks, and alter and change the terrain so much, that those who once lived there may not be able to locate the very spot their house once resided.

In the Philippines, on Southern Leyte, on February 17, 2006, a huge combination rockslide-mud-debris mass from nearby Mount Can-abag devastated the village of Guinsaugon. It caused widespread damage and a death toll of 1,126. This happened after a minor earthquake of magnitude 2.6 on the Richter scale, and battering torrential rains dumped 79" of rain in just 10 days! That is 45" more rain than my native city,

Milwaukee, receives in a whole year! So deep and wide-ranging was the landslide that those lucky few who happened to be away from the town when the mountain gave way, not only could not find their homes, but were not sure where much of the village had once stood. Many bodies have never been recovered.

What happened to the village of Guinsaugon would have occurred, in a much greater scale, worldwide, during the deluge. Guinsaugon was buried by relatively little in the way of rock, dirt and mud, and yet it just about disappeared. After Noah's raging flood you would not see a familiar sight, town, or city. Nor would there be a single recognizable landmark on the surface of this planet, for everything would be covered with thousands of feet of sediment. All traces of every original streambed, river, lake, and ocean would have been wiped away and covered with sediment. All hills, mountains, and plains would have been eroded, scoured of all vegetation and soil, which would then have been deposited as sediment over the entire surface of the earth. During the thousands of years since, they have hardened into sedimentary rock.

One other thing to consider, which we will briefly touch upon later, is that Noah's home would no longer be in the same location as when he boarded the Ark, for the continents shifted, moved hundreds, maybe thousands of miles from their previous locations.

THOUSANDS OF FEET OF DEFORMED SEDIMENTS

As you drive through road-cuts, or view photos of mountains, you will notice the layers of sedimentary rock, top to bottom, are deformed in a manner consistent with unconsolidated, soft, putty-like material that did not have time to harden before its formation. The layers of sediments are not millions of years old, nor is there a difference in age between the layers. All the layers were compressed and distorted at the same time. If the layers on the bottom were millions of years older than the subsequent deposits, would they not they have hardened into rock during those long eons of time, and cracked and broke when folded?

Dr. Morris believes that the whole column of strata, such as we find in the Grand Canyon, was formed at the same time, within hours or days. He stated this:

> Not only do such sedimentary rocks abound all over the world, but they give much evidence of having been formed by rapid and continuous depositional processes. Each individual stratum is a distinct sedimentary unit and, in most formations, can be shown by hydraulic analysis to have been formed within a few minutes' time. Furthermore, it can be shown that within a series of "conformable" strata, each subsequent stratum began to be deposited immediately after the preceding one. When the strata above and below a given interface are *not* conformable (such a surface is called an "unconformity" by geologists), then a significant time gap is indicated. However, since there are no *worldwide* unconformities, one can always find a place at which any given formation does grade conformably and imperceptibly into another formation above it, *without* a time gap. (emphasis his)[9]

As observed, when layers of sediments are first deposited they are always flat, horizontal, and in most cases parallel in relationship to each other. With the exception of most volcanoes, mountains are comprised of large portions of (sea-life) fossil-bearing sedimentary rock. Mountains are always composed (take a close look at any pictures of mountains) of twisted, buckled, folded-over, and bent layers that were squeezed together by lateral compression forces, not by an uplifting force, as many textbooks suggest. They are like a blanket that was folded many times and consequently buckled, as though it was pushed or squeezed together from both ends. The mountain building process which caused, what we observe in all of Earth's mountain ranges, was a onetime event and does not occur today. Present day processes, such as earthquakes and other tectonic activity do not form what we observe.

"Folded Mountain, British Colombia, Canada" by artiste9999
(istockphoto.com) photo ID:484565423

What is very interesting about those deformed and crumpled layers of sediments now hardened into rock is that there is little sign of breakage, cracking, or heat. Rock is brittle and will break before it will bend, but those layers are deformed like putty. Dr. Walt Brown[10] says this about major mountain ranges:

> Many of us have seen, especially in mountains and road cuts, thinly layered rocks that have been folded like a doubled-over phone book. How could brittle rock, showing little evidence of heating or cracking, fold? Sometimes these "bent" rocks are small enough to hold in one's hand... They must have been squeezed and folded soon after the sediments were laid down, but before they hardened chemically. But what squeezed and folded them?[11]

All of the Earth's current mountain ranges were once relatively flat ground upon which layers of sediments piled up. After this accumulation took place, it was then that the mountains were formed and the sediments hardened. This information corroborates the fact that not a single existing mountain range was present in the antediluvian world. Thus, there were no excessively high mountains that needed to be submerged during the flood. It is also the case that all mountain chains were formed at the same time. Since the flood, nowhere on earth have we found any disposition of sediments with creatures entombed in them on such a massive scale. In fact, nothing remotely resembling such layering is being created anywhere on earth today.

If the earth was nearly spherical, all smoothed out, with no valleys, mountains, or ocean trenches, there is enough water to cover the earth to a depth of about 8,000 ft, or a mile and a half deep.[12] This information clearly answers the charge of the ignorant, that there was not enough water to cover the tallest mountain peaks on Earth.

WHERE DID THE THOUSANDS OF FEET OF SEDIMENTARY ROCK COME FROM?

We have been discussing sedimentary rock in the last few pages, but have not proposed a solution as to where all those massive amounts of sediments came from that formed the thousands of feet of sedimentary rock. These strata were forming all over the world at the same time, so where did this eroded material, the new sediments that were piled on top of the old, come from?

Evolutionists never seem to consider this quandary. So, is there really a problem? Well, let us take a look. About 40 identified rock layers form the walls of the Grand Canyon. Evolutionism would have us believe that those sediments and materials—which seemingly came from nowhere—kept piling up, layer upon layer, without eroding or disturbing the previous layers. That somehow billions of creatures died, settled upon the ground, waited around for numerous years without decomposing, before being snugly tucked under tons of sediments. Evolutionism has a huge unrecognized problem with these numerous layers of sediments: Not only can they not explain where those sediments came from, but they have to explain where each different kind of sediment that composes the column came from. According to evolutionists this remarkable worldwide accomplishment took millions and millions of years to achieve. Each layer was created after deposition of sedimentation for millions of years; the process stopped for an indeterminable amount of time, and then started up again, at least 40 times, without disturbing the previous layers. Each succeeding layer added many feet of new sediments without eating away or absorbing material from the old layers.

On December 9, 2015, *Thrillist* featured an article titled "9 Things You Didn't Know about the Grand Canyon." A bit into the article is this caption: "The Great Unconformity will make your brain hurt." In this section is a picture of a man's hand touching two layers of rock on one of the walls of the Grand Canyon.

> See this guy's hand? It's touching a geological division called The Great Unconformity—the difference in age between the rock touching his upward fingers and the rock touching his thumb is more than 1 billion years. Let's explain... the Grand Canyon is really old... geologists have found that there are about a billion years of mysteriously missing geological history. Basically, the rocks found underneath the super-old rocks in the canyon don't date back as far. Scientists don't really agree on how it was formed. Today, more research is being done to understand the anomaly.[13]

However, what they believe does not fit with what they observe. There is no gap, or missing billion years of time (the evidence for a gap is missing), nor is this an anomaly. They only believe it is an *unconformity* because of their faith in evolutionism, not evidence or science. These geologists wrongly assume something, and then, based on that false assumption, do more research to understand how something happened when there is nothing to find, for there is no evidence that it ever occurred. Sorry guys, it is a waste of time. Reindeer don't fly.

Now, granted, they have not as yet found dinosaur fossils in the Grand Canyon, but in numerous other places, not only have they found fish, but land-dwelling animals and fowl. Are we to conclude that T-rex, along with his relatives and their prey, took a dip together in the ocean with birds and other beasts and all drowned together? After all, those sediments are for the most part water-laid deposits.

A possible solution as to where all that sedimentation came from lies in the longest mountain range on Earth, one many are not even aware of. This phenomenon is mostly hidden from prying eyes, for 90 percent of it lies deep under the waters of the earth's oceans. It is called the mid-ocean ridge, and stretches around the globe like the seam of a baseball. Its great extent was not fully known until the ocean floor was surveyed in detail during the 1950s. The entire formation is 49,700 miles long, and has a continuous system of 40,400 miles, making it greater in length than all other mountain ranges combined. It runs south through the middle of the Greenland Sea and the North and South Atlantic Oceans, then extends eastward into the Indian Ocean basin and onward between Australia and Antarctica, into the great Pacific Ocean basin. It continues northward along the eastern side of the Pacific basin. This chain of mountains is also the largest volcanic feature on the Earth, consisting of thousands of individual volcanic ridge segments that periodically erupt.

This somewhat narrow, globe-encircling series of mountains appear to be the Tectonic Plate boundaries. This is probably the place where the earth split open and allowed the flood waters to gush forth and cover the entire globe. It also is where the huge amount of sediments came from that formed the earth's immense layers of sedimentary rock.

DROWNED FROM BENEATH—THE HYDROPLATE THEORY

Some of you are probably thinking, it was 40 days of rain that caused the Flood of Noah, so what has water coming out of the earth got to do with anything related to the Flood? Let us ruminate for a bit about that great event, and consider earth sciences in relationship to the Flood and what Scripture states about it.

The Flood was not a series of miracles, with God intervening, suspending, or changing the laws of science to allow the deluge to take place. The Flood waters were not supernaturally created for this event, nor did they magically dissipate or evaporate into thin air when the disaster ended. The natural laws of science, stratigraphy, geology, and physics did not cease to function or operate out of the ordinary during the cataclysm. Their behavior, interactions, and dynamics were not altered to accommodate the events that took place. The flood did not supersede the laws and properties God engineered into matter in the beginning, which all nature conforms and adheres to. The only thing providential concerning the Flood might have been the timing. Then again, God knew how events would transpire on earth concerning the wickedness of man. So, in the beginning, maybe He synchronized the events and calibrated the timing, and just like a time bomb that has been set in motion, counting down to go off at the appointed time. In this case it was set for the eighth day after Noah and his family had entered the Ark.

However, speculations and questions such as these have no end, nor any answer we can be sure is the correct reason, solution, or response. The only thing we know for sure is that these events happened as recorded by Noah, and left an abundance of evidence all over the earth.

Someone did some calculations they thought proved the flood was not possible, for even if it did rain for 40 days and 40 nights, if all the moisture in the atmosphere precipitated out as rain, there would only be enough water to cover the ground 2" deep. This person could very well be correct, but therein lays the problem—when you start with the wrong assumptions you will certainly not come to the correct conclusions. *The rain was not the cause of the flood!* It was the result and direct consequence of the cause.

Let us read what Scripture relates in Genesis 7:11. Notice what happens first, and then occurs afterwards. *"In the six hundredth year of Noah's life, in the second month, the seventeenth day of the month, the same day were all the fountains of the great deep broken up, and the windows of heaven were opened.*" The flood began with the fountains of the deep breaking open (Gen. 7:11, 8:2; see also Pro. 8:28). Carl Baugh states this about the fountains of the great deep:

> This brings into play the second great natural force: *moving water*. Calculations have been made, demonstrating that major disruption of the "fountains of the great deep" below Earth's granite crust would send jets of steamy water ten to seventy miles high. These jets of steam would be sufficient to rip the canopy and open "windows" in its crystalline lattice, collapsing its structure in the forty-day timeline given in Scripture. The greater volume of water, however, would be expunged from the internal "fountains of the great deep," along with vast deposits of internal elements. This eruption of elements and water continued for 150 days, providing material and mechanism for the global sedimentary deposits.[14]

Dr. Walt Brown has developed the *hydroplate theory*, which, although not accepted by some within geological circles, seems to offer convincing evidence for the formation of many of this world's physical features. He believes the crust of the earth was 60 miles[15] thick. Beneath this layer were huge interconnected subterranean chambers or reservoirs of water that, because of the earth's crust pressing down on them, were under great pressure (about 10 million lbs per ft^2). Then, for some unknown reason, the pressure began to increase within the chambers and the crust cracked. The cause will never be certain; maybe it was tidal forces stressing the crust until it cracked, or a large asteroid struck the earth with devastating consequences. We will follow Dr. Brown's scenario starting from the cracking of the crust:

> ... The crack followed the path of least resistance ... circling the earth in several hours ... As the crack raced around the earth, the ten-mile-thick "roof" of overlying rock opened like a rip in a tightly stretched cloth ... Water exploded with

great violence out of the ten-mile-deep "slit," which wrapped around the earth.

All along this globe-circling rupture, a fountain of water jetted supersonically into and above the atmosphere. The water fragmented into an "ocean" of droplets that fell to the earth great distances away. This produced torrential rains such as the earth has never experienced-before or after...

The extreme force of the 46,000-mile-long sheet of upward jetting water rapidly eroded both sides of the crack. Eroded particles (or sediments) were swept up in the waters that gushed out from the rupture, giving the water a thick, muddy consistency. These sediments settled out over the earth's surface in days, trapping and burying many plants and animals, beginning the process of forming most of the world's fossils.

The rising flood waters eventually blanketed the water jetting from the rupture, although water still surged out of the rupture. Global flooding occurred over the earth's relatively smooth topography, since today's major mountains had not yet formed.[16]

What would it have been like, to have been one of the unfortunate antediluvians who failed to heed Noah's warnings, on that dreadful judgment day, when the covering of the great abyss was broken up and violently spewed out its contents? *"On this day were broken up all the fountains of the great abyss and the windows of the heaven were opened"* (Genesis 7:11).[17]

THE ANNIHILATION OF LORD LAMECH AND HIS ESTATE

The noble's citadel was built on a hill, and well-fortified to help protect it from the aggressive warring tribes that occasionally found their way through Havilah's impenetrable forests. From that vantage point, his sentinels could survey the entire landscape and keep an eye out for trouble. His realm, though created by violence and force of arms, was a place of plenty, filled with fruitful orchards; air perfumed by carpets of flowers, and terraced gardens brimming over with the best produce this side of Eden. There was however, of late, something seemingly wrong in his small kingdom. An unspoken uneasiness, a feeling of doom had settled upon his well-paid mercenaries, household servants, and slaves. Some were so terrified that they had even stolen away and left for safer places, far from this plagued land. All had heard rumors and stories, and many had been eyewitnesses of strange events. Small lakes and large streams drying up overnight, and pools of hot, bubbling, foul-smelling liquids oozing out of cracks in the hills of the northern pasture lands. Periodically throughout the day and night were heard muffled, deep rumbling sounds; at other times, it was a loud groaning, like something under great stress, which emanated from some undisclosed, undefined source. Lighting strikes, electrical discharges, and odd-colored glowing clouds and mists were seen in the evenings amongst the surrounding hills, where new fissures and cracks seemed to develop daily. Most disturbing was the disappearance of all wildlife. Huge flocks of birds, swarms of insects, and masses of animals had been seen fleeing, prey and predator together, heading south, and an eerie silence now enveloped the forest. Even the fish seemed to have vanished from the surrounding streams.

There had also been talk going around the compound, about some crazy man, Noah Haha, as most called him, preaching about a coming flood, a judgment from an angry God. He even had built an enormous ship, which those who believed would be able to board, and so save themselves and their families. "Perhaps," said some, "these strange occurrences and incidents might be a warning?" Although all were frightened about the recent manifestations and some thought about Noah's preaching, no one took heed.

The lord's herdsmen were also on edge, and very concerned with the master's prized aurochs. Normally even-tempered, unflinching, and pleasant to work with, lately they were easily spooked, as a growing agitation and restlessness spread among the herds. Their concern came to fruition when, startled by a sudden shutter from the ground, they went berserk, stampeded and broke through the surrounding fences, trampling two luckless stable hands, then ran straight to a nearby lake, plowed into the water, swam around in circles until exhausted, and then drowned. Out of 911 head, not a single one survived.

Lord Lamech was fatigued from the day's activities, so punishment for his herdsmen would have to wait until the morrow. He did, however, contemplate their fate... hanging... yes, all 23 would pay for their carelessness with their lives. Disposing of their bodies would not be a problem; his dragons would eat well tomorrow. Then, ignoring the other events and feeling the need to adhere to his normal routine, just as he always had, he chose one of his courtesans to spend the waning hours of daylight with. Today, Zillah received the favor of his company; while a slave served them intoxicating liqueurs and fruit fresh from the orchards; they snuggled together on his veranda and enjoyed another enchanting sunset.

Startled by a sudden lurching, heaving movement from the ground beneath them, they stole a fleeting terrified glance at each other, just before they were hurled to the ground. As the earth continued to pitch and heave violently there was a massive blast, then, a piercing, thunderous hissing sound that grew louder with each passing second. From his prone position, Lamech looked in the direction from which the sound seemed to emanate. In the distance, maybe ten miles away, he saw a dark rippling curtain of water jet up from a fissure in the ground, slice through the clouds and spew into the upper atmosphere. As far as he could see in either direction, this was taking place. Along with the water, even at that distance, he could see huge chunks of land, rock, and dirt being torn apart, pulverized, and quickly erode the rapidly expanding crack, which was then carried upwards with the water.

Gravity cannot be thwarted for long, and the slurry of eroded materials flung nearly into space soon started to disperse and separate into an ocean of grit-filled muddy droplets, and begin their long journey back down to the earth. As the horrifying incessant roar seemed to grow even louder, the sky quickly began to darken, but before it turned dark as midnight, he could see the wall of water continue to devour the land, and rapidly advance toward his estate; he knew there was no escape. In the gathering darkness he reached out, as he touched the trembling hand of Zillah and pulled her close, he felt the first large drops of grit filled rain pelt his face...just as his manor was swallowed up..."*It is a fearful thing to fall into the hands of the living God*" (Hebrews 10:31).

SUMMATION

What is described in this scenario, the eroding away of the walls of a 60-mile-thick, 46,000-mile-long rift in the earth for five months, would certainly account for the thick blanket of sediment we find that covers the entire earth. Let us read another portion of the flood account from the Scriptures:

> ...I will cause to rain upon the earth forty days and forty nights: and every living substance that I have made will I destroy from off the face of the earth...[12]And the rain was upon the earth forty days and forty nights...[17]And the flood was forty days upon the earth; and bare up the ark, and it was lift up above the earth. [18]And the waters prevailed and were increased greatly upon the earth; and the ark went upon the face of the waters... [19]And the waters prevailed exceedingly upon the earth; and all the high hills, that were under the whole heaven, were covered. [20]Fifteen cubits upwards did the waters prevail; and the mountains were covered. [24]And the waters prevailed upon the earth an hundred and fifty days.
> 8: [1]And God remembered Noah, and every living thing, and all the cattle that was with him in the ark: and God made a wind to pass over the earth, and the waters assuaged. [2]The

> fountains also of the deep and the windows of heaven were stopped, and the rain from heaven was restrained; ³And the waters returned from off the earth continually: and after the end of the hundred and fifty days the waters were abated. (Genesis 7:4–8:3)

When it rains hard and long enough to cause flooding, as soon as the rain stops the water will quickly start to drain (how quickly, will all depend on the terrain) from the flooded areas unless there is a continual source of water that feeds and continues to flow into the flooded area. If you read closely, the account of the Flood of Noah, you will find that the rain continued for just 40 days and nights. However, the waters continued to rise for 110 more days, even after the rain ceased! How is that possible? The Scriptural account lends credence to Dr. Brown's *hydroplate theory*. As Dr. Brown stated: "The rising flood waters eventually blanketed the water jetting from the rupture, although water still surged out of the rupture."

The rain stopped after 40 days, for its source was not, as many suppose, the clouds, but the water jetting into the atmosphere from the rupture. Once the flood waters were deep enough to quell its upward force and momentum, the rain ceased, for its source closed. However, although the rain had stopped, the water continued to pour forth from the abyss for almost four more months, until *"the fountains also of the deep and the windows of heaven were stopped . . . after the end of the hundred and fifty days the waters were abated"* (Genesis 8:3). For a total of five months the flood waters kept rising. Once they ceased, it took another seven months for the waters to recede, drain off the land, and settle into their new basins and beds. In total, for 371 days (53 weeks), Noah and his family lived aboard the Ark before disembarking.

In the Book of Psalms, we find what many believe is a reference to the receding of the waters after the flood:

> Who laid the foundations of the earth, that it should not be removed for ever. Thou coveredst it with the deep as with a

> garment; the waters stood above the mountains. At thy rebuke they fled; at the voice of thy thunder they hasted away. They go up by the mountains; they go down by the valleys unto the place which thou hast founded for them. Thou hast set a bound that they may not pass over; that they turn not again to cover the earth. (Psalm 104:5–9)

The mountain-forming process that Dr. Brown covered in some detail in his book must have begun toward the end of the Flood and continued for a very short time, less than a day. In fact, the evidence suggests that it occurred rapidly, within one hour or so.

Walt Brown has put the eighth edition (2008) of his book on his web site for the public to view. He asks this question in the book: "Could earth's mountain ranges form in less than an hour?" Here is part of his answer:

> An Analogy. Imagine that a long, massive train lost its brakes and is steadily gaining speed (accelerating) as it races down a high mountain. Eventually, this runaway train will crash. Its many boxcars will suddenly decelerate, compress, crush, and jackknife. In this analogy, the mountain the train is racing down represents the steep slope from the upbuckled Mid-Atlantic Ridge down to the subsided Pacific hydroplate; the tipped and crushed boxcars represent today's compressed and buckled mountain ranges . . .
>
> How rapidly did earth's mountains form? Although our runaway train picked up speed slowly, if, after hours of acceleration, its wheels suddenly fell off, the train would rapidly decelerate and crash. Likewise, the compression event began after most of the Atlantic had opened up. The hydroplates began to meet major resistances, ran out of lubricating water, decelerated, crushed, and buckled. Earth's mountain ranges were pushed up in less than an hour—all with fossils of sea life on top.[18]

The timing of this event coincides with the wind (Genesis 8:1) that Noah mentions. As the mountains were being rapidly lifted, this would have displaced the "ocean" of air residing over the continents, forcing it to flow swiftly away (as wind).

During and after that great mountain-building event, I am sure Noah and his family must have felt the earth tremble and settle, as Earth strove to attain equilibrium and stability after untold amounts of sediments were dumped over its entire surface and its subterranean chambers and reservoirs were emptied of most of its liquids. Numerous earthquakes and shaking probably continued for decades, hundreds of years after this tumultuous event. Bryan Nickel, a mechanical engineer, gives a detailed and impressive overview of Dr. Brown's hydroplate theory. You can find this six-part video on YouTube.[19]

Our home, Earth, is still largely a "water" planet. The huge oceans, mountain ranges, and numerous layers of fossil-bearing sedimentary rock is all a continual reminder, a testimony to a God who judges sin. Hopefully, in this chapter we have given a satisfactory answer to a few questions the reader may have entertained, or at the very least, given them a little insight, some light to help illuminate their path in the search for understanding.

We did not touch upon the water canopy that many suppose existed above the earth before the flood: We will discuss that in the next chapter.

NOTES ON CHAPTER 13

1. "Sedimentary Rock Types," *Encyclopaedia Britannica* web site, n.d., https://www.britannica.com/science/sedimentary-rock/Sedimentary rock, paras. 6.
2. Dan Nosowitz, "How to Become a Fossil After You Die," Atlas Obscura web site, May 01, 2015, http://www.atlasobscura.com/articles/how-to-become-a-fossil-after-you-die, paras. 17, 19.
3. "Perfectly Preserved 13,000-Year-Old Skeleton of Girl Discovered Deep in Underwater Cavern Hints How Humans Came to the Americas," *National Post* online edition, May 15, 2014, http://nationalpost.com/news/perfectly-preserved-13000-year-old-skeleton-of-girl-discovered-deep-in-underwater-

cavern-hints-how-humans-came-to-the-americas.
4. Vince Stricherz, "Ancient Bison Teeth Provide Window on Past Great Plains Climate, Vegetation," University of Washington News web site, August 7, 2006, http://www.washington.edu/news/2006/08/07/ancient-bison-teeth-provide-window -on-past-great-plains-climate-vegetation/
5. "Bison History," Texas Bison Association web site, n.d., http:// www. texasbison.org/bison_history.htm.
6. "New Research Confirms Origin of Bison in North America," Laboratory Equipment web site, March 14, 2017, https://www. laboratoryequipment.com/news/2017/03/new-research-confirms -origin-bison-north-america
7. Stuart Buchanan, "What an Ancient Wildebeest Teaches Us about Fossils and Evolution," Maropeng, the Official Visitor Centre of the Cradle of Humankind and the Sterkfontein Caves web site, February 26, 2016, http://www.maropeng.co.za/news/entry/what-an-ancient-wildebeest-teaches-us-about-fossils-and-evolution.
8. "Sedimentary Rock Formations of the Grand Canyon," In the Playground of Giants web site, n.d., http://intheplaygroundofgiants.com/geology-of-the-grand-canyon-region/sedimentary-rock-formations-of-the-grand-canyon-region/, para. 1.
9. Morris, 1976, p. 204.
10. Dr. Walt Brown is Director of the Center for Scientific Creation, a retired full colonel (Air Force), a West Point graduate, and has a Ph.D. in mechanical engineering from Massachusetts Institute of Technology. Dr. Brown has taught college courses in mathematics, physics, and computer science. Chief of Science and Technology Studies at the Air War College, Associate Professor at the U.S. Air Force Academy, and Director of Benet Research Development and Engineering Laboratories in Albany, New York.
11. Walt Brown, *In the Beginning: Compelling Evidence for Creation and the Flood* (Phoenix, AZ: Center for Scientific Creation, 2008) p. 80.
12. John Hudson Tiner, "Do We Have Enough Water to Flood the Whole Earth?" Creation Science 4 Kids web site, August 13, 2013, https://creationscience-4kids.com/do-we-have-enough-water-to-flood-the-whole-earth/
...Now to the depth of the water during the flood. Dr. Morris says that it has been calculated that on a smooth spherical earth the water would cover the earth to a depth of 8,000 ft. That is easy enough to check (provided I remember to use radius of earth rather than diameter!)
Radius of earth in miles: 3,963 miles
Thickness of 8,000 ft of water in miles: 8,000 ft/5280 ft/mi = 1.5 miles
Volume in cu mi of a smooth, spherical earth with a 1.5 mile thick layer of water:
$(4/3)pi*r^3 = (4/3)pi(3,963 + 1.5)^3 = 261,008,000,000$ (rounded to nearest million)
Volume in cu mi of a smooth, spherical earth without the layer of water:
$(4/3)pi*r^3 = (4/3)pi(3,963)^3 = 260,711,000,000$ (rounded to million)

13. The difference (261,008,000,000 – 260,711,000,000) is the of water to make a layer of water 1.5 miles thick: about million cu mi of water. (para. 7)
14. Lauren Topor, "9 Things You Didn't Know About the Grand Canyon," *Thrillist*, December 9, 2015, https://www.yahoo.com/lifestyle/9-things-you-didn-39-t-know-1317015596761142.html, para. 7.
15. Carl E. Baugh, "Crystalline Canopy Theory," Creation Evidence Museum of Texas, n.d., www.creationevidence.org/evidence/ crystillane_canopy_theory.php, para. 3. Carl E. Baugh is the Founder and Director of the Creation Evidence Museum of Texas in Glen Rose. He is the scientific research director of the world's first hyperbaric biosphere, simulating earth's atmospheric conditions before the world-wide flood of Noah's day.
16. Brown, 2008. In Dr. Brown's latest revision, he now believes the crust might have been closer to 60 miles thick, rather than 10 miles.
17. Ibid., p. 88.
18. Joseph Magil, The Englishman's Hebrew-English Old Testament Genesis—2 Samuel (Grand Rapids, MI: Zondervan, 1974).
19. Brown, 2008, Frequently Asked Questions: Could Earth's Mountain Ranges Form in Less Than an Hour?
20. Brian Nickel, "Hydroplate Theory Overview, Parts 1–6 Combined, Updated" [film], 2:41:12, August 8, 2016, https://www. youtube. com/ watch?v=4h-hE6tzJR_c.

> It cannot be denied that from a strictly philosophical standpoint geologists are here arguing in a circle. The succession of organisms has been determined by a study of their remains embedded in the rocks, and the relative ages of the rocks are determined by the remains of organisms that they contain.*

CHAPTER 14

THE WORLD THAT PERISHED

Have you ever wondered why God, instead of uprooting and destroying the wonderful and brilliantly designed environment of biomes and hydroponic systems with a flood, didn't just send a plague to wipe out the evil inhabitants on this planet, to which Noah and his family were given immunity? One reason might be the horrible stench caused by multiplied, millions and millions of dead antediluvians, their corpses rotting in their homes, yards and places of business. Another plausible reason against a plague type of judgment would be the unrestrained and uninhibited growth rate of the wild animals.

This entire world was to be man's home, shared with the animals. The millions of antediluvians acted as a restraint and hedge against the unfettered growth of the animal kingdom, but with all of mankind suddenly removed, except Noah and his family, the whole world would have been swarming with wild beasts of every sort, and would have outnumbered and eventually overwhelmed the small band of survivors. This was not a problem at first, for in the beginning all nature was at peace, but over time the effects of sin began to spread among the animal

---

*R. H. Rastall (Lecturer in Economic Geology, Cambridge University), *Encyclopaedia Britannica*, 1956, vol. 10, p. 168.

population. This was one reason why God did not allow Israel, at first, to conquer all the Promised Land. They were not yet sufficiently numerous to fill the whole land occupied by the wicked Canaanites.

> And I will...drive out the Hivite, the Canaanite, and the Hittite, from before thee. I will not drive them out from before thee in one year; lest the land become desolate, and the beast of the field multiply against thee. By little and little I will drive them out from before thee, until thou be increased, and inherit the land. (Exodus: 23:28-30) (See also Deut. 7:22)

However, I think the reason goes much deeper, something almost intangible, more abstract, a subtle and elusive cause. Such a plague could have wiped out all humanity, but a few hundred years after its devastation, what traces would remain of God's judgment on the world? In God's dealing with the Israelites, often He commanded them to make memorials, a reminder for their children, such as the time they crossed the Jordan River into the Promised Land:

> And Joshua said unto them, Pass over before the ark of the LORD your God into the midst of Jordan, and take ye up every man of you a stone upon his shoulder, according unto the number of the tribes of the children of Israel: That this may be a sign among you, that when in time to come, saying, What mean ye by these stones? Then ye shall answer them, That the waters of Jordan were cut off before the ark of the covenant of the LORD; when it passed over Jordan...and these stones shall be for a memorial unto the children of Israel for ever. (Joshua 4:5–7)

After a plague, ruins of decaying cities and villages would still dot the landscape here and there, but nothing much different than archeologists find and uncover today—certainly not a sure sign of judgment.

I think there are two main reasons: First, God brought the Flood, not just to destroy man and beast, but also the works of man, his civilization, his knowledge, and his technology. Without a total destruction of man's creations, his idols, places of ill-repute, where licentiousness and debauchery took place, sophisticated weapons of warfare, and advanced types of technology would still exist, just laying all round, waiting for someone to come along and pick it up and put it to use. In the four-episode miniseries *The Stand*, which aired on May 8–12, 1994, an apocalyptic drama based on the book by Stephen King, my suggested scenario took place. In a government laboratory a weaponized version of influenza is accidentally released. Containment proved futile, and the super-flu, nicknamed "Captain Trips" spread unchecked and caused civilization to collapse when it killed over 99 percent of the entire world's population in less than two months. The remaining survivors coalesced into two camps: one, led by kindly Mother Abagail, made a permanent settlement in Boulder, Colorado, while the other, headed by the demonic Randal Flagg, set up an autocratic society headquartered in Las Vegas, Nevada. Flagg wanted to destroy the opposition, so he decided to appropriate a nuclear weapon to wipe them out. He sent one of his minions to retrieve the device. However, before the missile could be deployed, a spectral hand reached out and detonated the bomb, destroying Las Vegas and apparently killing Flagg.

The other reason is that there is now worldwide evidence (an everlasting memorial), literally under the feet of all that are born, of Holy God's judgment upon the depravity, immorality, wantonness and impiety of rebellious men. Now men are *without excuse* and *willfully ignorant* if they do not believe.

DINOSAUR EXTINCTION THE RESULT OF A METEOR IMPACT?

Let's now consider the reasons ancient creatures have not reclaimed their former magnificence. Evolutionists have neither idea nor inkling as to why—but they do have a proposed cause of their extinction.

The evolutionary answer is that there have been at least five mass extinction events throughout Earth's history; some sources state that mass extinctions have occurred at least 13 times during the course of history. During each occurrence, much of life on earth was wiped out and needed to evolve again, only to be wiped out once more in the next mass extinction. This cycle has supposedly happened over and over during the course of millions of years. However, the main game-changer (which wiped out an estimated 70 percent of life) was the most recent one, which caused the mass extinction of the dinosaurs, happened 65 million years ago at the end of the Cretaceous period. At that time, supposedly no large land animals survived, tropical marine life was decimated, and vegetation was also greatly affected.

So what was the cause, the reason for the demise of much of the life on the planet? Not every evolutionist concurs, but at this time the main consensus is that a 6- to 9-mile-wide asteroid, traveling at 40,000 mph, slammed into the Earth near Chicxulub in Mexico and killed most living things. This event is sometimes called the K-T event. According to those who put forth this idea (physicist Luis Alvarez and his son, geologist Walter Alvarez), the explosion from an object this size would have released as much energy as 100 trillion tons of TNT, which is a billion times more than the atom bombs that destroyed Hiroshima and Nagasaki in 1945. It is true that an impact of that size and magnitude would most certainly cause the death of great numbers of animals, and great destruction of the flora over the entire earth. But would that blast leave the kind of evidence we now find which covers the entire earth? The intense heat of such a large impact would cause everything, probably within hundreds of miles of the impact crater, to be instantly incinerated, and everything outside of the incineration zone for untold distances would be torn to pieces and strewn all around and left to rot, unburied upon the ground.[1] Kind of like the Mount Saint Helens eruption, but much more destructive by many orders of magnitude. Would an event like that create numerous sedimentary rock layers filled with fossils? Would it cause sediments to lay down neatly on top of each other, with little or no evidence of erosion between them? In fact, can any process, any activity or method put forth by evolutionists concerning the numerous sedimentary rock layers covering the entire earth, explain what is observed? Nothing but a world-wide flood gives a satisfactory and adequate answer.

Some suggest that the K-T event produced a worldwide layer, about 1 cm thick, filled with a high concentration of the rare element iridium. Now that may be true, but as stated, that one-time event does not explain what we observe, over a mile of sedimentary rock layers covering the entire earth. The extinction of those great creatures was not caused by a large meteor impact, for any asteroid big enough to kill all the dinosaurs that inhabited every known part of the earth would most certainly have killed almost every other living thing on the planet, terrestrial and aquatic. Nor would a water impact from a large meteor cause the strata found worldwide to form, as evidenced by observed recent tsunamis. With a meteor strike you would get lots of dead animals, but, as Ken Ham, the founder of *Answers in Genesis*, says, it would not cause "Billions of dead things, buried in rock layers, laid down by water, all over the earth." Only a worldwide flood could account for that. Yet evolutionists continue to insist that there is no evidence for a world-wide flood, for the only evidence we have found are "billions of dead things, buried in rock layers, laid down by water all over the earth..."

Knowledgeable scientists and geologists who believe in the Scriptures know that the great extinction event that destroyed that marvelous world was not millions of years ago, nor was it caused by an enormous rock striking the earth—it took place about 4,500 years ago. And it was water, a worldwide flood, which separated the world that perished from the world that now is.

WHY AND HOW WERE CONDITIONS DIFFERENT?

Both camps, those folks who place their faith in evolutionism and those scientists who know that God created all things would probably agree on this point: something was vastly different. Before we consider a possible explanation for the different conditions on this present world compared to its ancient biosphere, let us lay out the genuine history of the Earth before the worldwide flood, an event that forever changed the ecosystems and meteorological conditions on this planet.

This Earth, solar system, and universe were all created within six literal 24-hour days, about 6,000 to 7,000 years ago. Everything was in perfect harmony with its surroundings, with a worldwide spring-like climate that allowed trees and other kinds of vegetation to grow at both poles! There were no scorching dry sandy deserts, or frozen barren wastelands. There were no Rockies, Andes, or Himalayas to climb, for all these massifs and chain of peaks were formed during and immediately after the flood. From creation to the flood, this virtual paradise lasted for over 1,600 years. Because of the favorable conditions prevailing on the Earth, man's and animals' lives were counted in centuries, not just years. Many reptiles such as snakes and crocodiles continue to grow all their lives. This is why many of those ancient beasts we call dinosaurs grew to such large proportions. Imagine an anaconda or crocodile living for 900 hundred or more years—what length and size could those reptiles attain? There is probably a biological limit, but before it was reached, as we have observed from the fossils, it allowed many creatures to grow to awesome sizes.

The dietary order of things in the beginning was different, because there were no carnivores. All animals, including lions and tigers, were vegetarian when first created. The animals were to eat green herbage and man was to eat the fruit and vegetables. You may think it is not possible that a lion or other kinds of carnivores could survive eating a diet of grass and vegetation as an ox or cow. However, there are true stories of modern-day carnivores that refuse to eat meat. There is an account of Little Tyke, the vegetarian African lioness who would spend an hour at a time eating tall succulent grass in the fields.[2]

At the present time, due to the Earth's axial tilt of 23.5°, most places on earth have four seasons—spring, summer, autumn, and winter.

However, the closer you get to the equator, the less pronounced they are. Here in the Philippines, the Filipinos say there are two seasons, summer and winter, but I do not believe it; for me winter means cold, snow, and ice. Here there are only two summer seasons—somewhat hot and very humid with a lot of rain, and very hot and not as humid with not as much rain.

Trees grew at both the north and south poles, confirming the whole earth was much warmer than today. How do we account for this, if, as today, each pole was shrouded in darkness for months at a time and gets intensely cold? The only seeming explanation is at the time of the flood something caused the earth to tilt to its present position or axial tilt of 23.5°. Before that time, the axis of earth was relatively perpendicular to the sun. This would mean that those who lived on the equator would have twelve hours of light and an equal amount of darkness year-round. However, the closer you got to the poles; the sunlight would become more diffuse and a lot less intense. If you were standing directly on one of the poles, each day, though the light would be dim, it would last longer, and the "night" would never get completely dark, for the brightness of the sun would "reach over" the top and bottom of the earth and cause there to be twilight for a distance on the other side of the earth.

There are other possible explanations, such as Dr. Brown's suggestion that the continents may have shifted during the flood due to the huge rupture in the earth, so much so that where semi-tropical plants and animals once flourished, the ground is now permafrost where trees do not grow.

Whatever the cause—whether the axis changed, the continents drifted, or a combination of both—something was vastly different. Let us see if we can uncover the reason, or at least a reasonable explanation as to why those ancient, antediluvian creatures were able to accomplish, size-wise, something very seldom equaled and never surpassed since.

We should first consider some of the ways things were different:

- Vegetation grew from pole to pole;
- The hydrologic cycle was vastly different;
- There were fairly uniform warm temperatures over the entire earth;
- There was more carbon dioxide in the atmosphere;
- There was a higher oxygen content;
- Insects, animals, fowl, marine creatures, and people were all much larger than they are now;
- People and animals had much longer lifespans;
- Earth's magnetic field was probably much stronger; and
- Air pressure was greater.

There is good evidence that numbers of so-called prehistoric animals and man were contemporary. Of course, the climate after the Flood was much different and many became extinct during the intervening years. However, some beasts survived for a time after their departure from the Ark, for they were able to adapt to the altered conditions on the earth. Unfortunately for animals such as the smilodon—better known as the saber-tooth tiger—the wooly mammoth, the mastodon, the elephant birds of Madagascar, the giant Teratorn, and others such as dragons, it seems that man played a large part in their demise. The largest known elephant, the *Steppe Mammoth*, was hunted by man to extinction. It grew to 15 ft in height and 10 to 12 tons in weight. Its great size was no protection against the intelligence of man, who used his wits and weapons to take down this impressive creature.

The first thing to consider on our list is the abundant and worldwide, pole-to-pole vegetation. Katia Moskvitch, in her article "Dinosaur Era Had 5 Times Today's CO2," said this about the great amount of carbon dioxide present during the time of the dinosaurs:

> Dinosaurs that roamed the Earth 250 million years ago knew a world with five times more carbon dioxide than is present on Earth today, researchers say ... "The higher CO2 levels [must] have [had] significant effects on the planet's climate, and its flora and fauna..."[3]

However, wouldn't more CO_2 in the atmosphere be a bad thing? It seems many have been brainwashed into believing the opposite of what's true. Contrary to climate alarmists such as Leonardo Wilhelm DiCaprio and Al Gore, CO_2 is not a villainous gas, a pollutant that will wreak havoc upon all life on earth. Plants need CO_2 for their photosynthesis, and the more there is, the better they will grow. It is the "miracle molecule of life,"[4] so it is needed to make the world a better and greener place, so breathe deeply and exhale more often. Higher CO_2 levels would certainly help explain why the vegetation was so lush, why it flourished and was so abundant. However, does that fact correlate, relate, or link together with the reason the animals were so large? Is there a cause and effect relationship between the two?

The soil was probably much more fertile, with all the trace minerals and nutrients in the proper amounts. There was a greater variety of foods and choices, so the animals' nutritional requirements were being meet. Thus, they were probably much healthier than many of their lesser ancestors. That much being assumed may give a partial answer, but really does not fully explain nor solve the mystery. The hydrologic cycle was drastically different in the antediluvian world. "*...the LORD God has not caused it to rain upon the earth...But there went up a mist from the earth, and watered the whole face of the ground*" (Genesis 2:6-7). It seems that aqueous vapors rising from the earth condensed in the colder regions of the atmosphere and fell back upon the earth in the form of fogs, dews, and mists, thus covering the plants and wetting the ground for absorption of water through the leaves and roots of plants. With this process rainwater would never leach away nutrients, nor would there be any erosion of the soil by heavy rainfall. There was much more CO_2, but that was not the only difference in the composition of the atmosphere. Based upon results obtained by USGS scientists from more than 300 analyses, they found oxygen levels were nearly 33 percent, compared to the present amount of 21 percent.[5] Anything beyond 35 percent oxygen in the atmosphere can create problems, but below that level is beneficial.

According to *Time Magazine*,[6] two scientists, geochemists Gary Landis of the U.S. Geological Survey and Robert Berner of Yale, were studying amber. They placed amber inside a vacuum chamber and cracked it open to analyze the trapped ancient air. They were startled to find their specimens contained about 50 percent more oxygen than it does now—about 32 percent oxygen, which is much higher than our current 21 percent content in the atmosphere.

Today most insects (arthropods) are relatively small (there are exceptions, such as the Japanese spider crab). Many breathe passively through special openings in their exoskeleton called spiracles, while others directly through their skin. Their growth and size is limited by their open circulatory system, which has no blood vessels to transport oxygenated cells throughout their bodies, and must wait for the oxygen to diffuse through their bodies. The diffusion of gases is effective over small distances but not over larger ones.

On the Prehistoric Wildlife web page is this information:

> It is generally considered that the maximum potential size of an insect is dictated by how much oxygen is available for respiration.

If there was more life-giving oxygen in the atmosphere of ancient earth, this would have given insects the ability to grow larger than their smaller and weaker descendants today. We thus have a probable answer as to why insects are not larger today.

While those results are not conclusive or irrefutable, if accurate, it would certainly help to explain, or at least place another piece of the puzzle on the board as to why everything grew to much larger proportions. But even an increase in the amount of oxygen in the atmosphere may not completely explain the reason how sauropod dinosaurs were able to take in enough oxygen to maintain their great sizes.

WHALE-SIZED SAUROPODS

Before we consider and ponder some of the problems concerning the immense size of ancient, pre-flood creatures such as the long-neck sauropods and others, let us reflect on this question: just how big did they get?

The largest and most complete (90 percent) and best-preserved Tyrannosaurus rex specimen (also the most expensive, after $7.6 million dollars was paid for it at auction) ever discovered is named Sue. It was named after paleontologist Susan Hendrickson, who discovered it on August 12, 1990, in the Cheyenne River Indian Reservation, South Dakota. This beast was 40 ft long, 12 ft tall at the hips, and was estimated to have weighed about 7 to 11 tons when alive. It now resides at the Field Museum of Natural History in Chicago, Illinois. Another large beast makes its home at the famed American Museum of Natural History in New York. On January 14, 2016 the skeleton cast (replica) of a 122-foot-long giant herbivore that belonged to a group known as titanosaurs was unveiled. Its estimated weight when alive was about 70 tons, which is as much as ten African elephants or 10 tons more than an M1 Abrams battle tank. These massive beasts have been unearthed on every continent.

Thanks to *Jurassic Park III*, the beast called Spinosaurus, supposed super-predator that vanquished the T-rex, is now somewhat famous. This croc-snouted, sail-backed theropod was supposedly a bit bigger and more menacing than the T-rex. Estimating its length at about 41 to 59 ft long would make it as large, if not larger, than the T rex. The first partial skeleton of this sail-backed beast was unearthed in 1912 by Richard Markgraaf in the Bahariya Formation of western Egypt. The original remains were destroyed in World War II, but at least six partial specimens have been discovered since. A lot of uncertainty remains, for there are missing elements of the skeleton that make any reconstruction, or ideas of how it walked, moved, or really looked, little more than conjecture. As with these remains, paleontologists recognize that there are multiple ways to interpret most fossils.

The big problem with most dinosaur finds such as the Spinosaurus is that it is an exceptional day when a complete skeleton is found. Most excavated skeletons are missing pieces, and even those that may be somewhat complete are broken, distorted, and damaged by the weight of the sediments that pressed down upon them. Sometimes the bones are mixed with those of other dinosaurs, or scattered over a large area. Another problem is trying to estimate the length of sauropods—most of the tail bones are missing, making it very difficult to come up with an accurate measurement. In an online article from *Scholastic*, dinosaur expert Don Lessem and paleontologists Tim Rowe and Bill Hammer answered some questions, including this one:

> Q: How many complete dinosaur bone sets have been found?
> A: Good question, and I'll be darned if I know the exact answer. One scientist estimated there are only about 2,100 good skeletons of any dinosaur in museums around the world. But a complete skeleton is another thing. It's not like a model kit that comes with all the parts included. When we are lucky enough to find whole dinosaurs it is usually because sand from a stream bottom or a sand dune has covered over the dinosaur soon after it died. But even then, the little bones of the tail are often washed or blown away. For instance, we have about 15 good skeletons of T. rex now, including two that are nearly complete. That's a lot compared to most dinosaurs, which are only known from a single tooth or bone. But we still don't have a complete T. rex.[8]

Was Don Lessem, our dinosaur expert, really serious when he said, "most dinosaurs... are only known from a single tooth or bone"? Well, yes he was. On July 30, 2016, Brian Switek's story "Paleo Profile: The Mystery Titanosaur" appeared in *Scientific American*. Yes sir, a caption stated: "A single bone gives away the existence of a previously unknown dinosaur in Brazil."[9] Only a single corroded, blemished fibula (one of the lower leg bones) was found. Let us hope they find the rest of the skeleton and reunite the fibula with its body, or for the remainder of eternity that beast will definitely walk with a limp. From that one bone, scientists estimated that he stood about 4.5 ft at the hip, and was not a fully-grown individual. Apparently his birth certificate was also found, for they dated his demise at 129 million years ago, give or take a year or two. Along with that document they must have found a family photo, for there was even a picture with the story, a speculative restoration of this beast, along with two other much larger specimens, presumably its father and mother. Just imagine what they would know if two or even three bones had been found.

So, any time you read about any extinct sauropod or theropod type of creature dug up by some paleontologist, or dinosaur "expert," just remember, any weight, height, or length, its age, or other so-called "facts" about those creatures should always be taken with a fair amount of skepticism and suspicion. Even if we could generally trust the opinions of the dinosaur "experts," there are many mysteries and questions surrounding sauropods, those fascinating long-necked, long-tailed creatures with blimp-like bodies. Elephants spend an average of 16 hours per day consuming 165 to 330 lbs of food. That's about 4 to 6 percent of their body weight! So how did a sauropod, which may have weighed 10 times as much as the average pachyderm, manage to find enough time to eat the required amount of food each day to stay alive?

BREATHE DEEPLY, LONG-NECK SAUROPODS

Another interesting problem is how did they get enough oxygen to their deep body tissues with small lungs, nostrils the size of a horse, and such long necks? With a neck that long, would it not be like breathing through a very long drinking straw or garden hose? Imagine breathing 24 hours a day through such a constricted tube. Wouldn't those beasts get a little red in the face trying not to suffocate (it was probably a problem because of the square-cube law, which we will discuss shortly)?

Would not the massive amount of air needed by such a huge animal, being sucked in and out, funneled through its small tube-like throat, heat up the air...? Well, now we know for sure: There were fire-breathing sauropods.

That also eliminates another problem, moving their huge bulk around, for all that hot air flowing into their lungs must have inflated them just like two hot air balloons, and that blimp-like body must have been mostly empty space waiting to be filled by the lungs from the hot air, giving them great buoyance. It is a wonder they did not just float away.

Some scientists say there is really not a problem, for the sauropods could breathe like birds:

> Furthermore, sauropods and other dinosaurs probably could breathe like birds, drawing fresh air through their lungs continuously, instead of having to breathe out before breathing in to fill their lungs with fresh air like mammals do. This may have helped sauropods get vital oxygen down their long necks to their lungs.[10]

Well, that might be true, but birds do not have to breathe through 30- or 40-foot long, tube-like necks! In fact, supposed sauropod experts, Michael Taylor and Mathew Wedel, estimated the neck of *Supersaurus* probably stretched out for 50 ft![11]

Another problem dealing with the tremendous length of the neck of the average sauropod is how did they hold their necks almost horizontally? The one we mentioned, the 122-foot-long herbivore, had a neck that stretched out for almost 40 ft! Try holding your arm horizontally for twenty minutes. It starts to get a little difficult after just a few minutes, now imagine your arm is 40 ft long. To solve this problem, some scientists have offered these solutions. "Sauropods had small, light heads that were easy to support."[12] Another secret was its mostly hollow neck bones, similar in make-up to the skeletal structure of birds. That might be a plausible or possible resolution.

There are other problems for these super-sized sauropods, and one consideration not thought about too often, is getting blood to their brains. Consider for a moment the giraffe. Because of its great height, up to 18 ft, a giraffe's blood pressure is far higher than that of any other animal. Such pressure would probable rupture the vascular system of most other animals. But God designed the pressure to be contained in such a way that it is maintained by thick arterial walls and very tight skin, which seems to function like a jet pilot's pressure suit. It is a good system, but if it ever failed when the giraffe bent down to get a drink, he would have a mind-blowing experience.

So, for a mature sauropod, when one held their head up, the blood might have to be pumped well over 50 ft in height! And when it bent down to quench its thirst, would not all that blood be under tremendous pressure, and rush to their head, pop its eyes out, and expand its brain just like a balloon? Well, no, for this creature's tremendous and marvelous system was engineered and designed by evolution, through happenstance, fluke, and quirk. However, under the present conditions on our planet, the atmospheric pressure and the oxygen content of the air, oversized insects and super-sized sauropods would not have been physically possible. For the same reasons, the large pterosaurs could not have flown in today's relatively thin atmosphere.

THE SQUARE-CUBE LAW

Another perplexing dilemma is how could these walking whales carry around that much weight? There is a mathematical principle, first described in 1638 by Galileo Galilei, called the square-cube law. It is applied in a variety of scientific fields, ranging from biomechanics to mechanical engineering. It states that as a shape grows in size, its volume grows faster than its surface area. It concerns the relationship between the volume and the area, as a body or mass's size increases or decreases. The square-cube law is the reason, for example, if you have a box kite that flies very well, and attempt to increase or decrease its size by a factor of 10, it will probably not be able to fly. In the Wikipedia article dealing with the square-cube law under the Biomechanics section it states:

> As was elucidated by J. B. S. Haldane, large animals do not look like small animals: an elephant cannot be mistaken for a mouse scaled up in size. This is due to allometric scaling: the bones of an elephant are necessarily proportionately much larger than the bones of a mouse, because they must carry proportionately higher weight. To quote from Haldane's seminal essay *On Being the Right Size*, "...consider a man 60 feet high...Giant Pope and Giant Pagan in the illustrated *Pilgrim's Progress*.... These monsters...weighed 1000 times as much as Christian. Every square inch of a giant bone had to support 10 times the weight borne by a square inch of human bone. As the human thigh-bone breaks under about 10 times the human weight, Pope and Pagan would have broken their thighs every time they took a step." Consequently, most animals show allometric scaling with increased size, both among species and within a species. The giant creatures seen in monster movies (e.g., Godzilla or King Kong) are also unrealistic, as their sheer size would force them to collapse.[13]

This is the reason the largest creatures to ever exist on earth are aquatic animals: To a great extent water negates the effects of gravity. Their musculoskeletal structures are mostly unhindered by the requirements of similarly sized land-dwelling creatures.

David Esker, college physics instructor, science researcher, and staunch evolutionist, discussed the square-cube law in his Dinosaur-Theory.com home page:

> ...Galileo's Square-Cube Law reveals what is wrong with the paleontologists' claim that there is nothing odd about dinosaurs and pterosaurs growing so large.

Over the years the evidence has become increasingly apparent as to why dinosaurs and pterosaurs should not be so large. For the large dinosaurs there are the problems of 1) insufficient bone strength, 2) insufficient muscle strength, and 3) unacceptably high blood pressure within the taller dinosaurs. For the pterosaurs there is the paradox that no cold-blooded reptile, not even the smallest reptiles, can fly today and yet during the Mesozoic era the cold-blooded pterosaurs grew to be the largest flying animals that ever existed. Clearly the belief that there is no scientific paradox regarding the exceptionally large dinosaurs and pterosaurs is incorrect. Yet the paleontology community has no means of saving face while backing down from their position, and so they continue to deny the paradox.

Esker also stated: "Once we understand Galileo's Square-Cube Law showing how size matters it becomes clear that the large dinosaurs and pterosaurs of the Mesozoic era present a scientific paradox." Esker believes he has come up with a solution to this real enigma. He has proposed a new scientific theory in an attempt to explain this phenomenon: *The Thick Atmosphere Solution*. He says his theory not only explains how those ancient creatures were able to attain astonishing sizes, but also solves the long-standing paleoclimatologist puzzle of how the same pleasant climate was maintained over the earth's entire surface. He believes that the air near the surface of the Earth was very dense, about two-thirds the density of water. Esker wrote: "In the same way that buoyancy enables whales to grow so large, the much thicker atmosphere provided an upward buoyancy force that reduced the effective weight of the dinosaurs, thus allowing them to grow to great size."

His theory includes the reasons, the primary mechanism by which the extremely thick atmosphere came into existence, and the explanation why today's atmosphere is much thinner. Esker's scientific curiosity drove his attempt to find a solution to the problem. While his theory is mostly nonsense because of its evolutionary accoutrements, a thicker atmosphere probably is the solution. Unlike most in the evolutionary community, he was honest enough to realize there was a problem and then do research to find an answer. It is just a shame that his honesty and sincerity did not reach just a little further.

From my perspective, it seems Esker did not arrive at his conclusion because of evolutionism, but despite it, for his solution for the most part appears to be based upon science, which evolution is not. This idea does not stand or fall upon evolution, but is independent of it. He does, however, attempt to tie it together with evolutionism, which creates a massive problem, because he starts with the belief that the solar system is 4.6 billion years old and assumes evolution (flying reindeer) happened. Starting with this faulty premise, assumption, and presupposition will never yield trustworthy or reliable results (his conclusion could not have come about because of his faulty premise). Using imaginary, fictional events and scenarios to base your experiments and studies upon will give you imaginary and fictional conclusions, and no matter how clever or inventive your research, it will turn out to be a fairy tale. Like the wonderfully imaginative fictional tales comprising the *Chronicles of Narnia* by C. S. Lewis and the *John Carter of Mars* stories by Edger Rice Burroughs, Esker's ideas are science fiction, just creative writing born of pure imagination.

THE CANOPY CONCEPT

So where do we turn to for an answer to this paradox? Esker's starting point was the square-cube law, a mathematical formula of real substance and usefulness. Unfortunately, for the rest of his research, he relied upon his religious faith (evolutionism) as his guideline and foundation upon which the rest of his doctrinal thesis stands.

Let us once again turn to the words of the author, architect, and designer of the universe's elements, its components, and the principles to which they must conform. The Creator's revelation was given to man, explaining events that transpired before man's creation. In Genesis 1:6–7 we read:

> And God said, Let there be a firmament in the midst of the waters, and let it divide the waters from the waters. And God made the firmament, and divided the waters which were under the firmament from the waters which were above the firmament: and it was so.

It seems from these verses and the context of chapter one, that the *firmament*, which in Hebrew means *expansion* or *spread out*, is apparently the atmosphere, the common abode of clouds and lofty soaring eagles. It was above this expanse that water, in its liquid, ice, or vapor state, was placed. As understood by many, this would have formed a canopy suspended above the earth. This next section will not be a thorough or comprehensive discussion of these verses. Nor will it be an in-depth study of the word *firmament*, but just a cursory review of the canopy theory. We will attempt to ascertain if this theory will solve or give adequate answers to our inquiry.

Down throughout the ages, various groups and cultures have believed that a dome or some kind of canopy covered the earth. Certain Mesopotamian cultures, such as the Babylonians, believed the sky was covered with a dome, firmament, or a solid vault. The ancient Chinese people believed the sky was a round dome, and the ancient Egyptians believed the sky was a ceiling or roof supported by pillars. In N.F. Gier's book *God, Reason, and the Evangelicals*, he stated:

> The final evidence I draw from rabbinic accounts. In Nachmanides' commentary on the Torah, he quotes from the ancient rabbis: "The heavens were in a fluid form on the first day, and on the second day they solidified." Another ancient rabbi said: "Let the firmament become like a plate, just as you say in Ex. 39:3." Nachmanides himself describes

> the firmament as "an extended substance congealed water separating" the waters from the waters. (15) Apart from the congealed water thesis, a modern Jewish Bible scholar agrees with this interpretation: "*raqia*" suggests a firm vault or dome over the earth. According to ancient belief, this vault which held the stars, provided the boundary beyond which the Divine dwelt.[14]

According to *Gesenius' Hebrew-Chaldee Lexicon* (#7549), the Hebrews believed there was a heavenly ocean of water beyond the stars, so it seems even the ancient Hebrews might have held to some of the beliefs and ideas of the oldest cultures. Because of this, some Christian's reason, the idea of a dome or canopy above the earth in the beginning must be myth. Some in the Creationist movement in recent times have discounted and begun to disregard the canopy concept. Even the *Answers in Genesis* ministry featured an article titled "The Collapse of the Canopy Model" by Bodie Hodge on its web site on September 25, 2009. Hodge concluded his piece with this statement:

> Answers in Genesis continues to encourage research and the development of scientific models. However, a good grasp of all biblical passages that are relevant to the topic must precede the scientific research and model. The canopy model may have a glimmer of hope still remaining, and I will leave that to the researchers, but both the biblical and scientific difficulties need to be addressed thoroughly.

Well-respected scientist, researcher, and hydraulic engineer, Henry M. Morris, together with John C. Whitcomb, Jr. wrote a book in 1961, *The Genesis Flood*, which is largely credited with starting the modern Creation Science movement. This well-known scientist believed there was a vapour canopy above the earth in the beginning. In Chapter One of his commentary, *The Genesis Record*, he stated:

> The "waters above the firmament" thus probably constituted a vast blanket of water vapour above the troposphere and possibly above the stratosphere as well, in the high-temperature region now known as the ionosphere, and extending far into space. They could not have been the clouds of water droplets which now float *in* the atmosphere, because the Scripture says they were "*above* the firmament." Furthermore, there was no "rain upon the earth" in those days (Genesis 2:5), nor any "bow in the cloud" (Genesis 9:13), both of which must have been present if these upper waters represented merely the regime of clouds which functions in the present hydrologic economy.[15]

Morris is not alone: There are many others who believe in and have developed canopy theories. In his online book *The Evolution of a Creationist: A Laymen's Guide to the Conflict between the Bible and Evolutionary Theory*, Dr. Jobe Martin stated:

> The giant flying reptiles such as the pterosaurs (pterodactyls and pteranodons) would be unable to fly in our present atmosphere. They needed a heavier atmosphere to get enough air to lift them with their 40-to 50-foot wingspans. Heaven-and-earth system #1 would have provided the heavier atmospheric pressure necessary for the flight of these huge creatures. Evolutionists say we don't know how these giant reptiles could have flown in our atmosphere. To a creationist, this is not a problem. Heaven-and-earth system #1, before the water canopy came down at the Flood of Noah's day, would have provided the air density needed for these huge creatures to fly.[16]

In the well-received book, *The Waters Above: Earth's Pre-Flood Vapor Canopy* (1982), Joseph C. Dillow continued the work of Dr. Henry Morris and John C. Whitcomb, authors of *The Genesis Flood*. He also discussed scientific evidence for a canopy and critiqued various canopy models. Of course, with any such look into the ancient earth, it is speculative at best.

The fact that this idea and a few other thoughts presented concerning how events, conditions, and circumstances may have operated or taken place in the past, may be a bit hypothetical, does not in any way cast doubt upon Scriptural text or its historicity. These are just suggestions, a reasonable attempt to use, at times, the sparse amount of information provided by reliable sources to fill in the gaps we have in our knowledge of the past, as done by others in all areas of research.

Carl Edward Baugh is the Founder and Director of the Creation Evidence Museum in Glen Rose, Texas. Although many in the Creation movement have distanced themselves from him because they consider some of his science spurious, this does not mean all of his scholarship is bogus. I have many books in my library, but I do not agree with every idea or thought those authors relay in their writings. However, I do not throw out the books, just because I may consider some of their beliefs incorrect. Baugh has developed a very detailed scientific treatise, something of substance, which he calls the *Crystalline Canopy Theory*. He stated:

> The firmament as a universal expanse, and its localized crystalline structure suspended above the Earth in the original pre-Flood creation, are viewed as one seamless whole in the concept and usage of the Hebrew word *raqiya*...

His model was peer-reviewed by Professor David Plaisted, PhD. After reviewing Baugh's theory and finding fault, and having deep reservations and issues concerning some of his science, Plaisted concluded:

> In general, there may have been a canopy around the earth at the creation. It may have had various benefits... There are and always will be questions about the nature of such a canopy, but in evaluating matters having to do with the creation, we should realize our limitations:

The secret things belong unto the Lord our God: but those things which are revealed belong unto us and to our children forever, that we may do all the words of this law." Only the Creator knows the specific details involved in the creation, but it is a gratifying pursuit to "think God's thoughts after Him."[17]

However flawed some of Baugh's science may be, he has constructed a hyperbaric biosphere where he will be conducting experiments. He will be using higher levels of oxygen and carbon dioxide at 2.18 Atmospheres of pressure in his biosphere, which he believes will simulate the pre-flood atmosphere and environment.

Dr. Kent Hovind has also developed a detailed canopy theory. He believes that the current 60 miles of atmosphere was squeezed into a super cold (-450F), perfectly clear crystalline ice canopy about 3" thick, suspended about ten miles above the ground. This 3" layer of ice would have weighed about 15 lbs per square foot, so 1.1 atmospheres would support it without needing the help of the magnetic field. However, he believes that ice at extremely low temperatures becomes laminated, metallic, and magnetic, and thus might have been held up by the earth's magnetic field, which was probably much stronger 6,000 years ago. You can check out all the particulars on Dr. Hovind's blog, "Dr. Kent Hovind Proves Canopy Theory Using the KJV Bible."

While dealing with the idea of some type of water canopy enclosing the Earth, a number of questions come to mind. If it worked so well in the beginning, why do away with it? The original was probably destroyed in the judgment at the time of the flood, and could not be replaced except by direct creation of God. He could have accomplished this, but the real reason He did not is known only to Him.

Hyperbaric oxygen treatment normally uses 100 percent oxygen for short periods of time. But would 31 percent oxygen, (higher than today) under higher air pressure for 100 percent of the time, yield the same effect? That is a question for the researchers to answer.

Although the authors mentioned above have differing opinions on just what the original canopy consisted of, most recognize that it was specifically designed to benefit planet Earth and its inhabitants. In the original creation, if there was a canopy of water above the atmosphere, solid, liquid, or vapor that enclosed the entire earth, what effects could this have had on the living creatures? And would this canopy provide some or all of the answers we have been searching for? A canopy located directly above the atmosphere would press down upon it, and cause the ambient air pressure to raise, in effect, life on the Earth would have been like living in a hyperbaric oxygen chamber. Consider the following.

Hyperbaric oxygen treatment (HBOT) is a relatively safe, non-invasive therapy. It is the medical use of oxygen at greater atmospheric pressure than normal, usually using 100 percent oxygen for short periods of time. Basically, it is just a way of delivering higher concentrations of oxygen under higher pressure. This results in higher levels of oxygen dissolving in the patient's blood plasma and deep body tissues.

On October 14, 1987, when Jessica McClure was only 18 months old, she was in her aunt's backyard and fell 22 ft below ground, down an 8" well casing. It took rescue workers working nonstop for 58 hours to free her. The story garnered international headlines. What some don't know about her rescue is that when she fell, her right leg was wedged alongside her body in such a way that her foot was next to her head. This odd position cut off circulation to her leg and part of her leg and foot began to blacken. Doctors wisely used HBOT, which helps the body fight off gangrene, and were able to save most of her leg and foot, restoring normal function except for one toe that had to be amputated. The rest of her foot was saved from amputation thanks to HBOT, which restores tissue oxygenation and helps reduce edema (tissue swelling).

A more recent case involving hyperbaric treatment occurred in February 2017. Thirty-one-year-old singer Meghan Linsey, a 2015 runner up on *The Voice*, was bitten on her face by a venomous spider, a brown recluse, found mostly in the southern, central, and Midwestern United States. She woke up to a stinging sensation on her face, and immediately went to urgent care (with the dead spider for identification). However, despite the use of saline and antibiotics, her left eye nearly swelled shut (antibiotics do not work for venom). During the next nine days she kept getting worse. Among other symptoms, she experienced muscle spasms, her entire body broke out in a rash, and she endured excruciating nerve pain in her face. On day nine, the symptoms lessened greatly, but necrosis (the death of cells in tissue or an organ caused by injury or disease) started on her face. She consulted with a wound specialist from South Carolina who placed her under hyperbaric treatment. After just three sessions, there were noticeably good results to what she called "a hole on [her] face" and she will not need surgery.

There are many hospitals and healthcare facilities across the country that now use cutting-edge HBOT technology. Among the prestigious institutes using this are American Cancer Society facilities, the New York-Presbyterian, one of the largest hospitals in the country, and the International Hyperbaric Medical Foundation. Many remarkable results have been achieved using this therapy. Some of the conditions that respond favorably to this treatment are cerebral palsy, autism, traumatic brain injuries, sports injuries, strokes, carbon monoxide poisoning, severe anemia, burns, hypoxic tissue, crushing injuries, skin grafts, radiation injuries, and heart attacks. Other areas of use include infections like diabetic gangrene, wounds that will not heal, ulcers, sepsis, pressure sores, and pneumonia.

So, living under a dome with hyperbaric conditions, how different would things be? How would it affect plant life, animal life, and the weather? This list contains a few of the ways life would probably have been different under the dome:

- Uniform world-wide warm temperatures.
- No rainfall, earth watered by mist, dew or fog (there was no rain at the time Adam was created. If there was no canopy, there might have been rain before the Flood).
- A water canopy would help filter out harmful rays such as ultraviolet radiation (UV) from the sun that damages skin and body cells.

- A thicker, heavier atmosphere would provide the requirements for flight for the larger flying reptiles.
- With uniformly higher atmospheric pressure, severe storms would be much milder, or impossible. With high pressure, we have clear, sunny skies, while low pressure brings rain, thunderstorms, tornados, and hurricanes.
- All living creatures, plant and animal, were much larger in stature.
- A canopy would provide the greater air pressure needed by giant insects, super-sized sauropods, and mammals so that oxygenated blood could reach deep into the body tissues and supply those tissues and organs with oxygen.
- With greater air pressure, small lungs would be able to sustain sauropods of tremendous sizes.
- Buoyancy supplied by the greater air pressure was needed by the large sauropods to help offset the effects of gravity;
- A much greater amount of oxygen and carbon dioxide could be contained in atmosphere;
- Most creatures decreased in size after the Flood;
- A greater longevity of life. The average age recorded before the Flood was 912; afterwards, the lifespan dropped off sharply for a few hundred years. Today few make it to 100.

Without a canopy, what mechanism, arrangement, or system is proposed by evolutionists or creationists to help the earth maintain the effects, benefits, and results listed, in the antediluvian world? Yes, there are a great many variables, and numerous difficulties in developing a working scientific computer model of the canopy theory. It seems to me to be somewhat arrogant, without knowing all of the variables that could come into play, to say the canopy model does not work—even if some of the science tested in a lab does not seem to work, this does not necessarily mean that the model would not work on a global scale.

In stating this, some may assume there is a flaw in my reasoning, for I reject the ideas proposed by a multitude of evolutionists because they cannot provide proof for every single aspect of their theories, and then insist that the canopy theory might be true, even though many struggle with their attempts to demonstrate it in computer models or simulations. Of course, some of the concepts of the canopy theory, as discussed, do work on a small scale, but would it be a viable theory for the entire Earth?

Contrariwise, as pointed out preciously, every aspect of evolutionism is based solely on faith, not science, has never been observed, you cannot test history, and you will not find proof for something which never happened. As evolutionists cling to their faith, I cannot be faulted for holding to a belief and faith in God, which throughout this book has been shown to be in accord with reason, science, and evidence, and reject evolutionism, for it is severely lacking these tenets.

The canopy theory is not a direct teaching of Scripture; it is only implied. While I do not want to be dogmatic, this much is certain—something was vastly different in the recent past. Conditions prevailed that allowed everything to attain enormous sizes and proportions that are unattainable under the present circumstances. Given what we know about the fossils and the past, it seems to me most of the problems, snags, and difficulties I have laid out in this chapter, are answered for the most part by the canopy theory.

NOTES ON CHAPTER 14

1. "K-T Event," JPL Public Information Office web site, n.d., https://www2.jpl.nasa.gov/sl9/back3.html
2. Little Tyke (1946–1955) was a female lion cub who was rescued by George and Margaret Westbeau. She refused all animal flesh from the time she was born. She refused to chew any bones and declined to drink any milk if there was even so much as one drop of blood in it. In a post about the grave memorial about this amazing lion, Kristy Arbuckle wrote: "Little Tyke, an African Lioness, was attacked by her mother at birth at the Point Defiance Zoo in Tacoma. She was nursed back to health by George and Margaret Westbeau of Auburn, Washington, and lived with them in their home for the rest of her life. She refused to eat meat, living instead on a diet of cooked cereals, eggs, milk, and cod liver oil. She was featured in many news reels, magazines, newspapers, television shows and at least one movie. Unfortunately, she contracted

a respiratory virus and died at age 9. Her owner George Westbeau wrote a book about her titled, *Little Tyke, The True Story of a Gentle Vegetarian Lioness* (1986).
3. Moskvitch, 2014, para. 1.
4. NaturalNews.com, Mike Adams, Carbon Dioxide revealed as the "Miracle Molecule of Life" for re-greening the planet, May 09, 2017.
5. "Air Bubbles, Amber, and Dinosaurs," USGS web site, March 4, 2013, http://minerals.cr.usg.gov/gips/na/amber.html; see also related information by John G. Cramer, "Dinosaur Breath," *Analog Science Fiction and Fact Magazine* web site, December 5, 1987, http://www.npl.washington.edu/av/altvw27.html
6. "Putting on Ancient Airs," *Time Magazine*, November 9, 1987.
7. "Meganeura," Prehistoric Wildlife web site, n.d., https://www.prehistoric-wildlife.com/species/m/meganeura.html
8. "Unearthing Dinosaur Bones and Fossils," Scholastic web site, n.d., https://www.scholastic.com/teachers/articles/teaching-content/unearthing-dinosaur-bones-and-fossils/
9. Brian Switek, "Paleo Profile: The Mystery Titanosaur," *Scientific American* July 30, 2016, https://blogs. scientificamerican. com/ laelaps/paleo-profile-the-mystery-titanosaur/
10. Charles Q. Choi, "How Dinosaurs Grew the World's Longest Necks," LiveScience web site, February 23, 2013, https://www.livescience. com/27376-how-dinosaurs-grew-longest-necks.html, para. 16.
11. "Dinosaur Reproduction, Not Ancient Gravity, Allowed Super-Sized Sauropods to Evolve," National Geographic Phenomena web site, February 25, 2013, http://phenomena. National geographic. com/ 2013/02/25/dinosaur-reproduction-not-ancient-gravity-made-sauropods-super-sized/
12. Ibid.
13. "Square-cube law," Wikipedia web site, September 29, 2017, https://en.wikipedia.org/wiki/Square%E2%80%93cube_law, para. 13.
14. *The Three-Story Universe*, From N. F. Gier, *God, Reason, and the Evangelicals* (University Press of America, 1987), chapter 13. http://www.webpages. uidaho.edu/ngier/gre13.htm, para. 13.
15. Morris, 1976, p. 59.
16. Jobe Martin, *The Evolution of a Creationist: A Layman's Guide to the Conflict between the Bible and Evolutionary Theory* (Rockwall, TX: Biblical Discipleship Publishers, 1994), p. 172.
17. David Plaisted, "Review of the Crystalline Canopy Theory," Creation Science Hall of Fame web site, http://creationsciencehalloffame.org/defenses/ark-flood/global-flood-peer-review/david-plaisted/crystalline-canopy-theory-review-david-plaisted/

> To suppose that the eye with all its inimitable contrivances for adjusting the focus to different distances, for admitting different amounts of light, and for the correction of spherical and chromatic aberration , could have been formed by natural selection, seems, I freely confess, absurd in the highest degree.*

CHAPTER 15

WHITTLE AWAY

MANY BELIEVE EVOLUTION WORKS like whittling, a little bit is removed from an organism, bit by bit, until what is left is able to produce a new feature, organ or ability. Of course, information would have to first reside within an organism before it could be whittled away—but how would removing information increase a creature's capabilities and produce new features and new organs with different functions and greater complexity? Then, after that new organ, feature, or ability was formed in the evolving creature and the process began again, a whittling away of information or features, how long could this process continue before the creature's data bank, its entire store of information, was completely drained? Or before the amount of DNA it contained was again equal to that found in its first ancestor, a rock? For when the process of evolution first began, it started with no information to whittle away.

Some believe evolution, because of external influences and pressures such as heat, cold, disease, environment, predation from carnivores acting upon an entire species or individual creatures, causes organisms to

---

*Charles Darwin in *The Origin of Species*, J.M.Dent & Sons Ltd, London, 1971, p. 167.

make small changes that are cumulative over a long period of time and produces new features or abilities in an organism. However, those kinds of pressures only bring out abilities and resources already possessed and resident within the animal. It does not cause the creature's data-bank to swell, nor gain new information, which would be needed to acquire additional features, characteristics, or new traits. Since that is so, what would be the process used by evolution to accumulate or expand information, or to infuse new data within an organism, from which said information could then be whittled away?

One evolutionist I know stated that Hemingway's polydactyl cats disprove this point. Polydactyly (also polydactylism or known as hyperdactyly) is a congenital physical anomaly (also observed in humans and dogs) that causes a cat to be born with more than the ordinary number of toes on one or more of its paws. This condition seems to be caused by a number of different mutations, and may not have come from the same line or from a single origin. I do not follow this line of evolutionary logic, that a mutation, a corruption of the genetic code of a feline would disprove any information offered. Now it may be true that radiation, pollution, or chemicals could harm the DNA and cause changes within. However, if that did occur, it would be a corruption of the original organism's genetic structure, causing deterioration and a weakening of the creature, not a building or creation of new information and betterment of the animal.

Information can be whittled away to create new traits or additional features? Try that with your computer when you want to make improvements or repairs. Whittle away a bit at some program, keep *randomly* removing information, letters, numbers or lines of code, a little here and a little there until you get new programs and new selections. Or just keep typing in *random* lines of code; add a little here and a little there, until you get new options, features, and settings. Remember, because evolution is a sluggish and unhurried process, you would not be spending hours each day pounding away at your keyboard typing in or removing indiscriminate selections. Every other week or so, you would drop by your computer and strike just a single random key. How many bad lines of code (mutations) would it take before your program crashed or became unusable? And how long would it take to fix it using evolution's trial and error method, randomly typing in numbers, letters or punctuation marks? If you said "Not a snowball's chance in hell," you nailed it.

"WHAT IS *TRULY* AMAZING ABOUT THIS IS THAT THE ARTIST KEPT *WHITTLING AWAY* WOOD FROM A *SINGLE TOOTHPICK* AND MANAGED TO CREATE THIS."

– Cartoon Studios –

When I was a machinist, at times I operated CNC machines. Every time the programmer created a new program for a part, if just a letter or two in a line of code was wrong, and you just went ahead and machined the part without a *dry-run* (ran the program through its cycle to observe that the correct moves were made without the chuck spinning and no part in the machine), what we called a *crash* might occur. I mean a literal crash—if the machine turret rotated and hit an 800 lb part that was spinning at a high rate of speed at the wrong time, you would not want to be anywhere near the machine. More than once, because of some mistake on the part of a programmer, the machinist's failure to catch his miscalculation, a failure of the machine, or other reasons, a large part left the confines of its enclosed space, and burst through some machines' doors as it tried to fly. Loud noises, flying machine doors, and large metal parts randomly moving through the air really gets the heart rate up, the adrenalin flowing, and a change of underwear necessary!

Compared to life, a machine program is extremely simple, very primitive, and yet just a few wrong letters or numbers in just one line of code out of hundreds or thousands of lines can create chaos. Thus it must be magic that produces evolution's needed changes, for there is no systems analyst to create new programs or functions, nor to work through innumerable lines of code to make needed changes, repairs, or corrections. Yes, animals do adapt to changes in their environment, but scientists have never observed this happen in any living organism: outside influences, pressures or chance that produces an *increase* in complexity in information in any living organism's DNA.

A recent news story[1] informs us that scientists are now able to manipulate DNA to enhance or augment certain aspects of an animal's physical characteristics, but that is not the same as generating or creating the information in the first place. All they managed to do was rearrange, reshuffle, and reorganize a smidgen of the animal's inconceivably complex DNA, which was made in the beginning by an intelligent being—God.

I. AUTOMOBILE

As we deal with this next section,[2] think about all the complex things, the steps and materials needed to make an automobile, and then contemplate that process, which, compared to a living organism is relatively simple. Is it easier to repair an existing automobile or to make a new one? To start from scratch with nothing but the dirt you are standing upon? Is it easier to heal a human body or to make a new one, forming it from the raw elements and materials found lying at your feet? There are many steps that must be taken in order to build a vehicle. For some procedures, any order of events will suffice. However, there are numerous procedures and other actions that must be taken in the proper order.

A car is created twice. Its birth takes place within the confines of an individual's mind. It starts with an idea, an image or concept. However, in reality there are very few original ideas. Many designs are just rearrangements of ideas previously conceived within the mind of another. A sculptor or artist does not come to the drawing table with an empty or blank slate, for his or her mind already contains a vast archive, a store of imagery and pictures to choose from, both manmade and natural. As he or she purposely muses about the subject at hand, a panorama of landscapes, tableaus and scenes flow before the mind's eye. The designer or architect simply builds upon the constructs and models already in existence. What is original is the way these ideas and thoughts are developed, organized, and assembled when touched by the sculptor's or artist's inspiration and creativity.

Once this first step has been taken, the actual physical creation or "birth" can begin. With an idea or concept now firmly planted within his mind, the next phase in the creation of the auto begins. Actual sketches, drawings, or models are created and fashioned, supplied by the images that reside in the designer's mind. Then, together with other artists, engineers, and draftsmen, machines will be retooled to make prototype parts and the process will continue until the vehicle comes off the assembly line.

Cars are made up of a variety of different metals, including aluminum, copper, titanium, and stainless steel. Numerous other materials such as leather, fabrics, plastics, and rubber are used in the finished product. There are different parts and systems, such as steering, suspension, exhaust, power train, engine, electrical wiring, wheels, chassis, and frame, along with numerous types of bolts, nuts, brackets, fasteners, wiring, etc., which go into the final product. The raw material needed to make steel and iron, the iron ore, must first be located by a trained geologist, who spent time being educated in his craft; the ore is then dug out of the ground by skilled miners. The trains used to transport the ore to factories and foundries must be designed and built. The ground on which the train tracks are laid must be surveyed, leveled, and prepared. The steel used for the tracks must be made and transported to the needed location and laid down by workers. People need to be trained to build, operate and maintain the trains.

The steel used in the manufacture of cars is made by removing most of the impurities from iron. Impurities such as sulfur, phosphorous, and silica weaken steel, so they must be eliminated. In the process of making steel, sometimes chemical cleaning agents called fluxes are used to help reduce the sulfur and phosphorous levels. Based on the chemical composition, steel can be categorized into four basic groups: carbon, alloy, stainless, and tool steels. By alloying steel with a variety of metals and chemicals, different properties can be attained. The addition of 10 to 30 percent chromium will create stainless steel. Stainless steel is usually divided into five types; depending upon the amounts of chromium, molybdenum, nitrogen, copper, niobium, aluminum, and various other elements, the finished product can be magnetic or non-magnetic, have different strengths, hardnesses, corrosion resistance, and toughness. Each batch of chemicals and materials added to the steel must be secured from the proper sources. Each ingredient must be manufactured to high standards, it must be graded, for each batch must be of uniform consistency, the same as those produced before and after. It must then be packaged or placed into containers and then labelled, shipped, or transported to the steel producer.

The glue used to affix the labels on the packaging must be invented, and then manufactured in factories built and designed for those proposes. Much research must go into the development of the chemicals and other materials used in creating the ink that is printed on the labels. The machines used in printing the labels must be designed and made. A graphic artist is needed to create and design the labels. An order must be placed by the customer, using any number of modern communication devices, phone, fax, or email, to order the labels. The order must be processed by the supplier and then shipped to the customer using various types of vehicles, all made by man in factories designed and built for those specific purposes and reasons.

Metals used for making electrical wires such as copper, aluminum, silver, or gold must be dug from the ground, processed, refined, and purified. The sheathing or insulating materials such as treated cloth or paper, the various oil-based products such as plastic, rubber-like polymers, or varnish the wires are wrapped or coated with must be made and applied. The wire may be further protected with substances like paraffin, preservative compounds such as bitumen, or may be wrapped with lead, aluminum sheathing, or steel taping.

Rubber trees need to be planted, fertilized, and pruned. The latex needs to be harvested by workers trained and able to make the crude rubber. It must be transported to the market, where a buyer will procure it and then sell it to the manufacturer, who will in turn make it into numerous products used in various applications, including the automobile.

Natural gas used for energy must be piped from the ground, using piping and materials designed and made for that purpose. Coal must be dug out of the earth by equipment engineered and built to supply the energy and power needed to manufacture the autos. The crude oil, the "black gold" used to make plastics, fuels, and numerous other products used in the manufacturing of automobiles, must be refined.

Before the development and use of safety glass, now commonly used in all autos, even a low-speed collision could cause gruesome injuries, for the broken jagged glass could easily slice through limbs, mutilating any passengers who were thrown around before the use of seatbelts, so safety glass is also needed. It was invented and designed not by accident, but by trained researchers, using skills and knowledge developed and accumulated over time.

The education of the workers, the scientists, and researchers that helped to create the auto was not an accident, nor done by chance. The paper in the textbooks used by those students was not printed by accident. The trees, where most paper comes, from did not invent nor make the chainsaw used to cut them down, nor do felled trees transport themselves to the sawmill. The chemicals used to make paper did not come about by chance. The employees working on the auto assembly line did not just happen to show up for work knowing what to do: They were trained, but not by accident or chance. Each individual process and stand-alone action is complete in itself, yet each separate composition is different from any of the other numerous processes and can be combined in a variety of ways to produce a bounty of outcomes and products.

The information listed above is just the "tip of the iceberg" concerning a few of the things that go into the manufacture of an automobile. There is no one on this earth, not a single person that understands, is knowledgeable or smart enough, to produce an automobile "from scratch" alone. No one understands every element of the whole process, the chain of events, or all the intimate, exhaustive details of transforming raw materials into a finished vehicle. All that knowledge resides within numerous people, and comes from a multitude of disciplines and sources from which someone had to choose, collect, organize, and apply if an auto was to be built.

There are countless acts of intellect, use of brainpower and aptitude behind every manmade product, none of which would have been made or accomplished without intelligence, wit, or cleverness. Many designs were accomplished only after observing the marvelously engineered living creatures designed by the master creator of all things—God. Each step for the automobile was thought out, purposely planned, executed, implemented, and accomplished by intelligent beings.

DESIGNED WITH PURPOSE

Thomas Edison invented the light bulb after numerous experiments and many long hours of research. Numerous materials he tried did not work. Though his process of finding the correct element was trial and error, it was purpose-driven and done intentionally. Similarly, his research to develop the alkaline storage battery involved over 10,000 experiments with different materials and chemicals before he was successful.

According to the authorized biography by Frank Dyer and T.C. Martin, *Edison: His Life and Inventions*, Edison's friend Walter S. Mallory observed that after an immense amount of thought and labor over more than five months of investigation and analysis, seven days a week, by his corps of chemists and experimenters, the research on the storage battery had achieved no results. Edison turned to him and replied: "Results! Why man, I have gotten lots of results! I know several thousand things that won't work!"

Ivory soap was one of the most unique and popular brands of soap in the world. It is incredibly pure (99.44 percent pure), produces a consistent and bubbly lather when used, and it floats on water (since 1992 they have sold a version that does not float). It floats because air is whipped into the soap mixture before it sets, making it slightly less dense than water, thus allowing it to float. For over a century it has been produced. But its originality, the unique ability to float on water, was claimed to have been created accidently by an unnamed employee who left a batch of soap mixture mix for too long, causing too much air to be introduced. That story was officially debunked by Ivory's own archivists in 2004: A notebook from 1863 was found in which a chemist working for Proctor and Gamble (the company that owns Ivory) noted he had made floating soap and thought all of their stock should be made that way.

What is really interesting about this story is that the discovery was made by James Gamble, the son of the man who founded the company, James Gamble Sr. Yet somehow the fact that he invented one of the company's most famous products has somehow been forgotten for over a century. However, even if the soap had been created accidently, the machines used in the manufacturing process were not. The power that ran the machines did not accidently flow to the factory. The buildings that housed the machines were designed and built by the hand of man. And the chemicals and other ingredients used in producing the soap were purified by man. The whole process of making Ivory soap was deliberate. Its launch in 1879 was purposeful, not accidental. It was achieved and carried out through the result of planning, with the intention, aim, objective, and motivation of making a profit from its sales.

There are a number of well-known products that were discovered unintentionally. Roy Plunkett, a chemist who worked for DuPont, accidentally stumbled across the non-reactive, non-stick chemical, one of the slipperiest substances known to man—Teflon. Post-It Notes, one of the most-purchased office products in the world, was discovered in 1968 by Spencer Silver, a chemist working for 3M when he stumbled across a low-tack adhesive while trying to make a super-strong adhesive for use in airplane manufacturing. Play-Doh, the modeling clay children have been playing with for decades, was originally used as wallpaper cleaner; at someone's suggestion, the detergent ingredient was taken out, an almond scent and coloring was added, and Play-Doh was created. Harry Coover, a researcher at Kodak Laboratories, stumbled across Super Glue while developing plastic lenses for gun sights. He abandoned his discovery, but nine years later, it was rediscovered—again by accident—by Fred Joyner, who chose not to abandon his discovery, and a commercial product was born. Other accidental discoveries include the artificial sweetener Saccharin, popsicles, plastic, the child's toy known as the Slinky, vulcanized rubber created by Charles Goodyear, and, if you want to nuke your food, you can use the microwave invented by Percy Spencer in 1945. All these things are manmade, but compared to any living thing they are simple and uncomplicated. In all cases, whether it was accidently discovered or purposely created, an immense amount of time, effort, and intelligence, as evidenced by the narrative of the auto, were required before it became a reality. The idea that something such as life, however, which is much more complicated by countless orders of magnitude, evolutionists assure us can create itself by quirk, happenstance, or a twist of fate. I fail to see the logic in that kind of reasoning.

INHERENT PROGRAMING

Water is called the *universal solvent*, for it is capable of dissolving more substances than any other liquid. Another of water's amazing properties is that, as it cools down, it contracts until it reaches 39°F and then expands, becoming less dense until it reaches the freezing point, thus giving ice greater volume so that it floats. If water did not have this remarkable feature, most life on this planet would not be possible. Ice actually acts as an insulator: It shields the water beneath it from the cold air above, and helps keep the water below from freezing. If this process suddenly ceased and no longer occurred, cold water, being denser, would settle to the bottom of any collection of water, and all lakes, rivers, and oceans in the far northern and extreme southern latitudes would begin to freeze solid each winter and kill most aquatic life. Each summer, the heat from the sun would only be able to melt the surface layer of ice in those latitudes and, like the plot out of a science fiction movie in which an experiment goes horribly wrong, eventually this frozen element would creep and spread until it engulfed the entire planet, making it a permanently frozen-solid wasteland, containing as much life as a popsicle.

Water's ability to expand in its solid form is inherent within its molecules. No increase or decrease in information takes place when the freezing process occurs. Likewise, when creatures or organisms adapt or change, no increase in information ever takes place. This adaptive ability, characteristic, and capacity already resides within and is fundamental and natural to all living creatures' genetic makeup. The ability to survive by adaptation and change does not come from surroundings or location; nor is it formed by any external environmental forces or factors. A bolt of lightning surges with immense energy and raw power, over 5 billion Joules' worth, and can reach temperatures of over 53,500 degrees Fahrenheit, hotter than the surface of the sun, but it lacks the ability to program or create a database or archive of information. It does not transfer software into what it strikes, nor does it have the ability to encode, place, form, or create information within the genetic material of any living organism. Likewise, any characteristics acquired during an organism's lifetime cannot be carried over to their offspring, for those things do not cause new programing or new information to take up residence within the genes or DNA. So, although the farmer's wife cut off the tails of the blind mice with a carving knife; assuming they were males that were married, when their wives conceive offspring, their babies will still be born with tails. Only what has been previously programmed and resides within an organism's genes and DNA will be available for use during its lifetime and for its offspring. According to evolutionists, given enough time, drugs cause bacteria to evolve, to mutate into different kinds of critters, better-adapted organisms that will keep on changing, evolving until they become something else, a new species with different features and capabilities they were not endowed with before; in time one of them may grow a head, arms, and legs, and learn how to twist along to a Chubby Checker song...wait just a minute, evolutionists already believe that happened. If that has supposedly already occurred, why could it not develop further, acquiring wings and evolving into an organic type of jet engine capable of flight at supersonic speeds?

PREPROGRAMED RESPONSE

Bodybuilders' training regimen, diet, supplements, and drugs taken during training will carry them only so far, for there is always a limiting factor, their inherited genetic makeup. The strain of lifting weights and the exertion and labor expended during weight training does not carry forward to their offspring. For bodybuilders, drugs act as a kind of fertilizer and nutrition for the cells of the muscles, stimulating, coaxing the cells, and prompting most systems of the body to work harder for optimal performance. However, any outside influences such as drugs, natural forces, or man's tampering, can never create new information within internal structures, DNA, or genes of any organism, because outside stimuli have no ability to impart new information or reprogram a cell.

A native of Egypt now residing in Massachusetts in the Boston suburb of Milford, bodybuilder Moustafa Ismail is sometimes called the Egyptian Popeye, for his eye-popping 31" biceps and triceps. In 2013 he was credited by Guinness World Records as having the largest upper-arm muscles on earth. However, some have pointed out that his forearms are way out of proportion to his biceps and triceps, as are his arms in comparison to the rest of his body. He looks so anatomically incorrect that many suspect, although he denies it, that he injects synthol or other additives directly into his upper arms to increase their girth. But the Guinness World Records has made it absolutely clear that it has seen no medical evidence to support any claim that Ismail used Synthol or any other body-enhancing substance.[3]

Depending on the criteria and whose list you use for the best bodybuilder, Ronnie Coleman, with eight Mr. Olympia titles, is claimed by many to be the biggest, most muscular human being to ever live. At 5 ft 11" tall, 340 lbs (300 contest weight), with a 60" chest, 36" thighs, 20" forearms, and 24" arms, he is a huge monster of a man. Supposedly able to deadlift over 800 lbs and using 180-lb dumbbells for curling reps, he is as strong as he is big. Gunter Schlierkamp would come in second on the body-builders Hall of Fame list; Markus Ruhl is number three; and Arnold Schwarzenegger, the most famous, seven-time winner of the Mr. Olympia title, is ranked, somewhat surprisingly, as only number six. The man who starred in the original TV series *The Hulk* as the title character, Lou Ferrigno, a 6'5", 285 lb colossus of a man, comes in at number seven. Most, if not all, of the top guns in men's and women's body-building have used or are still using steroids or different kinds of drugs to make their muscles grow larger, sometimes to amazing, almost inhuman sizes, shapes, and dimensions, which at one time were found only in the realm of fantasy and cartoon renderings. But, I ask, is it possible that the physiques of Moustafa Ismail or Ronnie Coleman, which were coaxed into excess by vigorous weight training and spurred on by possible drug use, could be passed down to their children? Are the mountains of muscle-mass acquired by top bodybuilders, something that has now become genetically inheritable or transferrable to their offspring? Not a chance!

Such a transference process has never been observed because it cannot happen; however, although it violates known scientific principles and laws, this does not stop evolutionists from teaching it, or science fiction authors from writing books filled with it, or Hollywood movie producers from making movies that present this nonsensical evolutionary idea as fact and science, something that can actually take place. The 2014 blockbuster movie *Lucy*, starring Scarlett Johansson and Morgan Freeman, opened at number one at the box office in the United States, Canada, and all other markets where it was released. It ranks at number 171 in terms of earnings to date, with a respectable worldwide total of $463.4 million. As with most examples of its genre, has an evolutionary-based theme.

Johansson plays a 25-year-old American woman living and studying in Taipei, Taiwan, named Lucy. She is forced to be a drug mule and has a pouch full of a synthetic drug sewn inside her abdomen. The movie's plot revolves upon the myth that we humans only use 10 percent of our brains, and this drug supposedly has the capacity to allow the user to unlock and tap into that unused 90 percent. Before Lucy is able to deliver the drug, she is beaten and the pouch inside her ruptures, flooding her body with the synthetic compound. As the drug kicks in, Lucy slowly becomes stronger mentally and physically as she is able to access more of her brain. As her power and capacities grow, she finds she has super Jedi-like powers of telepathy and telekinesis, and can control people and animals with her mind. As her abilities begin to soar off the charts, she becomes super-human like the members of the X-Men gang; she feels no pain and is able to tap into and control any type of electronic device. As the movie concludes, she finally reaches full control of 100 percent of her brain, where her powers have grown to the point that her body begins to metamorphose into a bizarre black shapeless substance and spreads over the computers and other objects in the lab she occupies. She transforms these things into the next generation supercomputer, which contains all of her enhanced knowledge of the universe. This knowledge is then downloaded to a compact black flash drive, which is offered by a tentacle-like arm of the advanced supercomputer to her friend Professor Norman (Morgan Freeman), just before she is transformed into a god-like spirit being and disintegrates to dust. When someone asks where Lucy is, immediately a text message appears on a cell phone: "*I am everywhere.*" Next we hear Lucy's voice stating: "Life was given to us a billion years ago. Now you know what to do with it." With that, the main point of life is made clear: to pass on knowledge.

Lucy's reaction to the drug is supposedly an evolutionary response, a speeded-up version that causes evolution to happen at a much accelerated rate. The drug allowed her to "crack open" a cell's nucleus, thus giving her access to her full potential, all the encoded data found in each cell in her brain (this movie is fiction but mirrors evolutionary thinking). This is where the assumption of evolution and faith in its process comes into play: it is a belief that life is always changing and moving to a higher plain, a more advanced, more improved, more evolved state, which means a more complex and developed organism.

The questions this movie or evolutionists never entertain, ask, or seek to explain, are where does the cell get the information (encoded into our DNA), the knowledge, data, and abilities Lucy was able to access? As outside forces such as drugs never impart programing or information, who or what programed the cell and imparted this incredible, astounding, and staggering amount of accessible information and data that was already residing within its genetic composition? The idea that cells contain information, stores of data, and huge reams of "programs" that can be accessed given the right set of circumstances is not science fiction. What is science fiction is the evolutionary idea that an outside force such as radiation or a drug can cause a cell to evolve and produce out of thin air, information, abilities, resources, and programming not already present in its genetic makeup. That somehow a cell is a mini-super computer that has the ability to conjure up and create programming, data, and information from out of nowhere.

Many traits possessed by creatures are hidden, concealed, suppressed, and buried deep within their DNA, but when some kind of stress or pressure is applied, these hidden capabilities, powers, and adaptabilities spring forth. By observing the same fish species that live aboveground in rivers of Mexico and Texas that have perfect vision, scientists who have been studying the blind Mexican cavefish think they have solved the riddle of why those fish lost their eyes in the permanent darkness. They hypothesize that, as eyesight is useless in total darkness, being sightless would give them an advantage when it comes to survival, because sacrificing that sense would allow more energy to be devoted to the hunt for food, which is very scarce in caves. The midbrain in these fish, the part that deals with vision, is significantly smaller, so the energy saving, accordingly, was 15 percent over a sighted fish, because their visual centers' energy-hungry neurons and photoreceptive cells were no longer active:

> One such process in Mexican blind cavefish is a phenomenon called pleiotropy, in which genes usually involved in eye

development are reassigned to features more useful to life in caves, such as increased numbers of taste cells for finding food in the dark.⁴

- Cartoon Studios -

Of course, this suggested reason for eye loss is still just a hypothesis, says William Jeffery, a biologist at the University of Maryland, College Park. He wrote in an email: "But while the new study 'confirms the feasibility'" of this eye-loss theory, "it doesn't provide compelling evidence."⁵ Further research shows that there may be multiple driving factors that cause genes to mutate, but it is really unknown how outside pressure "tampers" with the fish's vision. Regardless, no matter what the cause of this supposedly evolutionary process, no increase in information is ever realized or attained. This process simply allows the vision genes to switch off or become inactive.

In another study[6] it was found that interbreeding the same species of sighted fish with blind cavefish will result, often within a single generation, their vision being restored. What occurs when this happens is that switch in the visual systems that was deactivated and turned off in the blind fish, was reactivated by the introduction of a sighted fish to the gene pool. Again, there was neither increase nor growth of information or complexity in the fish species, just a reactivation and rebirth of an existing, but non-functional or dormant vision program.

ACCESS GRANTED TO REAMS OF KNOWLEDGE

Most animals do things by instinct, abilities and knowledge that are inherent rather than learned. They are born with this instinct fully developed and active in their genetic makeup. These abilities are not formed, learned, or acquired over long periods of time by trial and error. Acquired characteristics, on the other hand, are never inherited. Things learned or information assimilated during one's lifetime cannot be infused or imbedded within the DNA or genes of an individual and passed down to their offspring. So where does this immense, hidden store of information contained within the DNA of all creatures come from?

Mankind is no stranger to these concealed and masked unknown abilities, some of which only emerge during times of stress and deep emotions caused by horrendous events in a person's life. A divorce or the death of a loved one will at times cause a person to draw deeply from a well of strength that seemingly appeared from out of nowhere. I am very familiar with a man who had never written a poem in his life, failed English in school, never finished high school, was plagued by depression for years after a trying divorce, distraught after the death of his young daughter and his mother, but was able to pen inspiring poetry after each one of those shattering occurrences. On April 16, 2015, PBS presented the NOVA program, *The Great Math Mystery*, in which they discussed the possibility that somehow mathematical abilities are part of the fabric, the very essence and structure that is built into the human brain. As you read the following, consider this: Who did the programing?

> To many mathematicians it feels like math is discovered rather than invented. But is that just a feeling, could it be that mathematics is purely a product of the human brain? ... *are the foundations of math built into our brains*? Even without any mathematical education, even without learning any number words or symbols, we would still have, all of us as humans, a primitive number sense, that fundamental ability to perceive number, seems to be a very important foundation and without it, it's very questionable as to whether we could ever appreciate symbolic mathematics, *the building blocks of mathematics may be preprogramed into our brains, part of the basic tool kit for survival.* (emphasis added)[7]

Occasionally some people develop amazing capabilities almost instantaneously, after a stroke, blow to the head, disabling injury, sickness, or traumatic accident seemingly rewires the circuitry of their brain. This new ability is just like accessing a new program on your computer. When you purchased your computer, it was preprogramed with a number of factory-installed features, numerous drives, programs, and databases. They were always there right from the start, and were just waiting for you to access and use them. The problem with the latent and covert programs nestled within man's brain is the ability to retrieve, open, or freely gain access to them for, unlike a computer; an instruction manual does not yet exist.

In certain respects, the human brain can be likened to a gasoline engine: Even when it is idling, each part is operational, fully functional, and being utilized. In man's brain, there seems to be in place a governor or rheostat that limits the power, hinders full throttle from being applied, but occasionally, when this governor is able to be breached and the rheostat overridden, there flow forth incredible achievements. In 1923 a case was reported in which, after contracting meningitis, a three-year-old developed musical genius. In 1980 there was a case of a nine-year-old who received a bullet wound to the left brain which caused muteness, deafness, and left-sided paralysis, but precipitated new mechanical skills. In 1991 there was a case of an eight-year-old who began to show exceptional calendar calculation skills following a left hemispherectomy. The condition known as *savant syndrome* is a clear case in point. Some of those folks afflicted with this are severely autistic, while others have Asperger's syndrome or some other kind of severe handicap. Though some of those affected have very limited, mental, social, and other life skills, some possess what Doctor Darold A. Treffert calls *islands of genius*, special skills or separate intelligences. These skills and capabilities are hidden away deep within the recesses of their mind, fully blown and fully functional but formerly dormant, buried, and not able to be accessed until the right set of circumstances happened to occur. Outside forces or stimuli do not form, generate, or create these capabilities. Outside pressures simply spur or induce human beings to access, develop and utilize what already exists.

On the web page *Listverse* is an extremely interesting article by Andrew Handley, "10 Fascinating People with Savant Syndrome." Below are a few short tidbits relating to three fascinating people.

Jason Padgett's ability did not manifest until he was 30 years of age, when he was given a very rare gift (there are only about 40 people who are known to have this), *acquired savant syndrome*. After being brutally mugged by two men who repeatedly kicked him in the head, when he awoke from his coma the next day he saw the world differently, through the lens of geometry. Everywhere he looked, he saw fractal patterns, mathematical designs hidden in objects, and distinct geometric shapes, and was able to draw stunning mathematically precise artwork that illustrated his intuitive understanding of complex mathematics. Somehow his brain had been able to make structural changes, rewire itself after being damaged, and was able to activate dormant areas of his mind, causing him to become a mathematical genius.

Tom Wiggins, although regarded as an autistic savant, was an amazing man. He was born of slave parents and blind from birth. In 1860 at age 14 he was the first African-American to give a command performance on a piano at the White House for President James Buchanan. He could play two pieces of music on the piano simultaneously with his right and left hand while singing another song. He could play a piece with his back to the piano and his hands inverted. He was also able to play perfectly, uncirculated piano compositions he only heard once. His memory was exceptional and he never forgot anything. Novelist Willa Cather, in an article for the *Nebraska State Journal*, called Tom "a human phonograph, a sort of animated memory, with sound producing power." He also had the remarkable ability to perfectly mimic nearly every sound he heard.

Leslie Lemke has savant syndrome. He is blind, has brain damage, cerebral palsy, and suffers from a few other maladies. Throughout much of his early childhood, was mostly unresponsive to external stimuli. Although never taught, he has innate access to all the *rules of music*. He has the ability to store and replicate music after only a single hearing. When he was just 14, was able to play back Tchaikovsky's First Piano Concerto flawlessly and can do that to this day on request. His learning level is that of a seven-year-old. He can, however, play nearly a dozen instruments and can recall perfectly every musical piece he has learned.[8]

All living organisms have *genetic memory*, a store of information that allows the transfer of knowledge and skills that are amassed but dormant, captive, just waiting for an opportunity to manifest—the flipping of a switch that will allow rewiring and release.

Dr. Treffert, MD, who currently practices at St. Agnes Hospital in Fond du Lac, Wisconsin, has spent 50 years studying and observing those affected with savant syndrome. In his article "Savant Syndrome 2013—Myths and Realities," he stated:

> Genetic memory—the genetic transfer of knowledge and skills—in my view accounts for the already stored dormant capacity tapped by the recruitment, rewiring and release... I base that concept on the fact that some savants, particularly those severely limited in other ways, clearly "know things they never learned." The only possible way to know things one never learned—sometimes at complex levels—is for that knowledge to be *factory installed, genetically transmitted.* (emphasis added)[9]

I love Dr. Treffert's descriptive terminology of knowing information never learned, which so eloquently fits the reality of what has been observed in savants, "*factory installed, genetically transmitte*d." In her article discussing Jason Padgett's abilities, "A Beautiful Mind: Brain Injury Turns Man into Math Genius," Tanya Lewis wrote:

> So do abilities like Padgett's lie dormant in everyone, waiting to be uncovered? [...]
>
> Most likely, there is something dormant in everyone that Padgett tapped into, Brogaard said. "It would be quite a coincidence if he were to have that particular special brain and then have an injury," she said. "And he's not the only [acquired savant]."
>
> In addition to head injuries, mental disease has also been known to reveal latent abilities. And Brogaard and others have done studies that suggest zapping the brains of normal people using TMS can temporarily bring out unusual mathematical and artistic skills.[10]

Padgett believes he is an example that everyone has untapped genius potential. In his memoir, *Struck by Genius: How a Brain Injury Made Me a Mathematical Marvel*, he stated "I believe I am living proof that these powers lie dormant in all of us...If it could happen to me, it could happen to anyone." All living organisms, human beings included, it seems from what we have read, come fully equipped from birth with innumerable energies, capabilities, and knowledge, but for most those resources will forever remain untapped. The abilities savants exhibit cover a wide gamut of skills and proficiencies including, but not limited to, musical genius, artistry (woodcarving, drawing, painting, sculpting), mechanics, exceptional calendar calculating skills, and astonishing memory and memorization abilities dealing with, among other things, music, birthdays, historical facts, sports trivia, bus or train schedules, maps and navigation abilities. Some of these skills and abilities can at times can be equaled but are rarely surpassed by "normal" people, or even by prodigies or geniuses. The skills exhibited can in some cases be learned; there are formulas for calendar calculating and memory techniques, such as those used by Harry Lorayne, Jerry Lucas and others. In the following quote, Dr. Treffert is referring to just one skill, but it applies across the board, to all skills and abilities exhibited by savants:

> …there are formulas for calendar calculating. And yes, if any person puts his or her mind to it, he or she can learn (laboriously) how to calendar calculate. But savants seem to have this algorithm or formula "unconsciously" inscribed or inculcated in their brain, and in most such individuals there simply has not been any study of the calendar nor the "learning" of any formula.[11]

The question we need to ask, which is never really addressed, is where did the unfathomable, immeasurable amount of latent abilities and information *factory-installed* in the brain, which was *genetically transmitted*, come from? Evolution's trial and error method can never account for this, nor provide us with the answer.

The complexity of the single-cell creature, the amoeba, is comparable to the intricacy of any cell within the human body. However, the human body as a whole is much more complicated, for numerous cells are combined into sophisticated systems and organs that function individually and collectively. The process of making an automobile involves the collection of innumerable ideas, designs, schemes, plans, and purposes—all which originated and flowed from the mind of man. All the wonders, technology, expertise, know-how, and skill that went into producing the marvels dreamed up by man took brainpower to accomplish. And yet, all of man's brilliant creations have a great deal less sophistication, intricacy, and complexity then an amoeba or any other living creature, and lack the most mysterious entity in the entire cosmos, *life*. Yet, somehow a lifeless and mindless nothing, evolution, is supposedly able to significantly surpass all of man's efforts, achievements and triumphs, with no intelligence, goals, plans, purpose, reasons, aspirations, objective, or direction. Truly, reindeer don't fly!

NOTES ON CHAPTER 15

1. Brad Reed, "Scientists Have Created a New Breed of Stronger, Faster Dogs Using DNA Manipulation," BGR web site, October 21, 2015, http://bgr.com/2015/10/20/china-dna-manipulation-stronger-dogs/
2. This section dealing with the creation of an automobile was inspired by the article titled "I, Pencil," which appeared in the November 1983 edition of *The Freeman*, 33(11).
3. Guinness World Records News, http://www.guinness worldrecords.com/news/, published January 1, 2013.
4. James Owen, "How this Cave-Dwelling Fish Lost Its Eyes to Evolution," *National Geographic* web site, September 11, 2015, http://news.nationalgeographic.com/2015/09/150911-blind-cavefish-animals-science-vision-evolution/, para. 19.
5. Brian Handwerk, "Blind Cavefish Can Produce Sighted Offspring," *National Geographic News*, January 8, 2008, http://news.national geographic.com/news/2008/01/080108-cave-fish.html, p. 2.
6. Ibid.
7. "Mathematics Behind Reality," More than Big web site, n.d., http://morethanbig.com/videos/watch/rq9iIX_Abnc
8. Andrew Handley, "10 Fascinating People with Savant Syndrome," Listverse web site, July 23, 2013, http://listverse.com/2013/07/23/10-fascinating-people-with-savant-syndrome/
9. Darold Treffert, "Savant Syndrome 2013—Myths and Realities," Wis-

consin Medical Society web site, https://www. wisconsinmedicalsociety.org/professional/savant-syndrome/resources/articles/savant-syndrome-2013-myths-and-realities/, para. 13.
10. Tanya Lewis, "A Beautiful Mind: Brain Injury Turns Man into Math Genius," LiveScience web site, May 5, 2014, https://www. livescience.com/45349-brain-injury-turns-man-into-math-genius.html, paras. 22–24.
11. Treffert, n.d., "Some mysteries remain," para. 16.

> The fossils that decorate our family tree are so scarce that there are still more scientists than specimens. The remarkable fact is that all the physical evidence we have for human evolution can still be placed, with room to spare, inside a single coffin!*

CHAPTER 16
ORIGINS AND THE TEACHER[1]

THE FOLLOWING CHAPTER IS a partial summation of the arguments, points, and issues presented so far. It is in a simplified form so that a person of normal intelligence, or even a Harvard professor educated in evolutionary doctrine, can understand it. The teachings of the faithful, the evolutionists, have not been distorted, deformed, misapplied, or misrepresented.

In most government schools, the following could well be an actual exchange between an intelligent and well-informed eighth-grade student and Mr. Smith, their honest but somewhat dippy biology teacher.

"Oh, Mr. Smith, could you deal with some questions I have concerning the origins of life?"

"Yes, Jane. What are your questions?"

"How old is the universe and the earth, and where did life come from?"

"Jane, since you asked, I am going to answer your questions about the age of the universe, the earth, and where you came from. Let's start with our great, great, great, great, great grandparents: We'll call them Mr. and Mrs. Rock."

---

*Dr Lyall Watson, 'The water people'. *Science Digest,* vol. 90, May 1982, p. 44.

"Mr. Smith, are these people?"

"No, they are rocks, just like granite and quartz."

"Rocks?"

"Yes, very intelligent rocks."

"Intelligent rocks?"

"Well, not really intelligent rocks, but rocks that must have contained information the first living thing used to make itself a long time ago when life first began."

"Mr. Smith, where did this information the rocks had come from, and how was it able to organize itself into something alive?"

"They don't know."

"Mr. Smith, is this science?"

"Well, no, but evolutionists like to call their faith science when it's really a religion.[2] Their religion teaches that all things created themselves and came into existence without the help of God. It is the belief in the supposed history of the earth; a story dealing with blind faith in things which cannot be repeated, in which experiments cannot be conducted, cannot be tested, cannot be measured, and have never been seen or observed. Without the ability to test whether a hypothesis[3] is true or false, and which can neither be confirmed nor falsified, evolutionism cannot be considered scientific, but is 'faith in things not seen.'

The belief in evolution actually violates at least two immutable laws of biology. One is the law of biogenesis, *life only comes from life*, and the other is *like always gives rise to like*. For it is obviously true that life always comes from other life, not inorganic matter, and that horses always give birth to horses, dogs give birth to puppies, lions always give rise to other lions, and humans always have babies that are human.

Of course, evolutionists have to believe, and have faith that the laws of science were different when life started and that the first living things could do, what science has shown that they cannot do today, or their story would only be a fairy tale.

Well, to continue... water ran off these rocks into a primordial sea, and some of the information that evolutionists have no clue as to the origin of, such as DNA, must have come off of the rocks, or maybe from somewhere else, because somehow, someway, once upon a time, life, maybe scum or a one-celled animal used the information that washed off of the rocks to organize, program and design itself into a living cell, which is something more infinitely complicated and complex than any computer or man-made marvel, and it became alive."

"Really! How?"

"There was probably lightning, and it hit the primordial soup and presto, life!"

"But don't living things die when they are hit by lightning today?"

"Well, yes, most of the time."

"So, Mr. Smith, how could it cause life way back then?"

"I don't know, but evolutionists somehow know these things, and they have faith in their beliefs."

"But Mr. Smith, is that what we see today, life forming itself out of inorganic matter such as dirt or rocks?"

"Well, no, but evolutionists assure us that spontaneous generation[4] did happen, once upon a time."

"But Mr. Smith, isn't the belief in spontaneous generation a belief that science abandoned many years ago after it was disproved by Francesco Redi in 1688, by Lazzaro Spallanzani in 1780, and again by Louis Pasteur in 1860?"

"Well, ah, yes. But now getting back to the rocks... let's talk about where the intelligence and 'magic rocks' came from. The most current and popular conjured-up story is it came from an explosion."

"What blew up?"

"Well, evolutionists do not how it got there, where it came from, how long it was there, why it exploded, and how all of the matter in the known universe could have compressed itself into a tiny speck the size of an atom, but they have faith that it was that minuscule particle of matter that exploded, or as they try to explain it, 'expanded like cake in an oven,' a sort of slow-motion explosion, that somehow created everything in the known universe."

"But an explosion?"

"Yes Jane, evolutionists try to explain the origin of the universe with what they call the Big Bang. However, their account really does not explain or address the cause for the universe. We know that everything that has a beginning has a cause. Evolutionists believe that the universe had a beginning; therefore the universe has a cause. But as for what caused the Big Bang, they do not have a clue."

"Okay, so the Big Bang does not answer the question of where the universe came from, as it does not address the cause of the Big Bang. But I am still puzzled about their belief. Don't explosions just break and destroy things, and turn organized structures into chaos and debris?"

"Yes, that is what we see and observe. But, Jane, this was a special one-time event."

"Oh. So when and where did this happen, and how long ago?"

"Well, in a galaxy far, far away, a long, long, long time ago, maybe a billion, or two billion, or 13.7 billion or maybe, ah . . . well, I do not think they know, because they keep rewriting their story. Then, after the explosion, the disorganized pieces of rock and matter somehow organized itself into stars, galaxies, solar systems, and planets with all of their precise movements and orbits."

"Well, Mr. Smith, after our solar system and our planet somehow made themselves, what did Earth's first life look like?"

"They don't know."

"How long before it evolved into something else?"

"They do not know."

"What did it eat?"

"They do not know."

"What do they know for sure?"

"They don't know."

"Oh. How long before frogs evolved?"

"Millions of years."

"What did they eat, and did flies evolve at the same time?"

"They do not know."

"Mr. Smith, how did chickens develop or originate? You can't get a chicken without an egg, and you cannot get an egg without a chicken."

"I told you, they do not know what they know."

"Mr. Smith, what does any of this have to do with science?"

"Nothing. As I told you before, it is a religious belief, not science. Evolutionists walk by faith, not evidence. Well, Jane, I hope I have answered your questions on how life started and where you came from."

"You didn't, but thank you for helping me to understand the religion that evolutionists believe and put their faith and trust in."[5]

NOTES ON CHAPTER 16

1. This chapter originally appeared in Chapter 6 of *Musings on Creation and Evolution*. It has been revised.
2. *Webster's Third New International Dictionary and Seven-Language Dictionary* defines "religion" as follows: "7 a: a cause, principle, system of tenets held with ardor, devotion, conscientiousness, and faith: a value held to be of supreme importance . . . and by practicing as well as preaching its doctrines."
3. *Webster's Third New International Dictionary and Seven Language Dictionary* defines "hypothesis" as follows: "1: a proposition tentatively assumed in order to draw out its logical or empirical consequences and so test its accord with facts that are known or may be determined . . . of such a nature as to be either proved or disproved by comparison with observed facts . . ."
4. Marshall and Sandra Hall explained spontaneous generation: "People once believed that new generations of living things arose from nonliving matter. For example, snakes were believed to arise from horsehairs and flies from decaying meat. This false idea about the production of living things was called spontaneous generation" (*The Truth: God or Evolution?* Direction Books, 1974, p. 17).
5. Michael Ruse is a leading evolutionist who recently published a book titled *The Evolution-Creation Struggle* (Cambridge: Harvard University Press, 2005) which deals with the "struggle for existence" between two rival religious systems. In its prologue, he stated:

    In particular, I argue that in both evolution and creation we have rival religious responses to a crisis of faith—rival stories of origins, rival judgments about the meaning of human life, rival sets of moral dictates, and above all what theologians call rival eschatologies—pictures of the future and of what lies ahead for humankind.

You will be greatly disappointed (by the forthcoming book); it will be grievously too hypothetical. It will very likely be of no other service than collocating some facts; though I myself think I see my way approximately on the origin of the species. But, alas, how frequent, how almost universal it is in an author to persuade himself of the truth of his own dogmas.*

CHAPTER 17

HOW LONG WILL YOU WAVER BETWEEN TWO OPINIONS?

Is not my word like as a fire? Saith the Lord; and like a hammer that breaketh the rock in pieces? (Jeremiah 23:29)

IT WAS ANOTHER SCORCHING hot day when Ahab, without his usual entourage of bodyguards and soldiers, took his servant Obadiah out of the city on a quest to search for water for the king's livestock. The king sent Obadiah in one direction, toward the formerly well-watered grasslands and marshes where wading birds once flourished; the king went in a different direction. He hoped that they would find some hidden source of water that had not as yet dried up. Suddenly, Elijah appeared and began to chastise the wayward king. He told him that because of the sins of his family, his people, and his own forsaking of the worship and service of the true God by serving false deities, that

---
*Charles Darwin, 1858, in a letter to a colleague regarding the concluding chapters of his Origin of Species. As quoted in 'John Lofton's Journal', *The Washington Times*, 8 February 1984.

it was Jehovah who had sent the drought that continued to plague the land. Boldly, he told the king to summon all his prophets and gather together all Israel unto Mount Carmel. Then, as suddenly as he had appeared, and before the king could object or fully react to the orders given by the intrepid prophet, he was gone. The king, weak-minded man that he was, dutifully submitted and ordered it to be done.

As the sun rose that morning, there was no foreseeable change in the weather forecast. As it had been for over three years, the descendants of Jacob knew it would be another cloudless day in the parched and dusty Land of Promise. But all of that was about to change, for on that very day atop Mount Carmel, a dispute was about to be settled. A clash between Jehovah's prophet, Elijah, and the prophets of Baal and Asherah was about to commence.

Three and a half years earlier, Elijah the Tishbite made a pronouncement to Ahab, the wicked king of Israel. He told Ahab that, without his word, there would be neither dew nor rain in Israel. Then, according to Elijah's declaration, God sent a drought as a testimony against the idolatry of Ahab and the people of Israel. God bestows the rain and sunshine on the just and unjust. Even though Israel had not yet repented of their wicked ways, God, in His mercy, had decided it was time for the rain to fall once again. However, before the rain would commence, there would be a showdown between the prophet of God and the 850 priests of Baal and Asherah. The king blamed Elijah for the lack of rain and wanted vengeance, but while the drought pitilessly seared the land, God kept Elijah safely hidden from the king's wrath until the appointed time.

On the appointed day, Elijah spoke these words to the gathered masses: "*How long will you waver between two opinions? If the LORD is God, follow him; but if Baal is God, follow him*" (I Kings 18:21 NIV). Then, to those assembled, he outlined how the events of the day would unfold. Wood and two bulls, one for the priests of Baal and one for Elijah, would be placed upon an altar and offered as a sacrifice. No fire would be lit under either offering. Baal's priests would proceed first to call upon their god. He then told the people *"the god who answers by fire, he is God"* (I Kings 18:24 NIV). Baal's prophets tried all morning to summon their god, to no avail. At noon, Elijah the prophet began to taunt the priests of Baal and Asherah. He asked them: *"Is your god sleeping or on a long journey? Or perhaps he is defecating, or maybe he is busy or in such deep thought that he cannot hear you."* "Shout louder," Elijah sneered. They did as he suggested and began to shout louder and continue their frantic gesticulations and strange behaviors.

While they called on the name of their god, they made hideous cries as they leaped upon the altar they had made. Morning till noon they shouted, while slashing themselves and one another with swords and spears until their blood flowed. Midday passed, and still no answer came, so they continued to dance violently around the altar. It was nearing the time for the evening sacrifice. Weary and exhausted, they ceased their useless antics. After the priests of Baal and Asherah failed to urge their gods to rain down fire upon the sacrifice they had prepared in their honor, it was Elijah's turn to call upon his God. He called the people together and used twelve stones, one stone for each tribe of Israel, to repair the broken altar of Jehovah. He then formed a large trench that encircled the altar. The wood and bullock were laid thereupon. He then told the people gathered to pour four barrels of water on top of the prepared offering. This was done three times until water filled the trench.

While the water still sloshed around in the trench, Elijah uttered a short prayer of 63 words to the God of Heaven. God's answer came as fire. It fell from the sky and devoured the burnt sacrifice and the wood; even the stones were consumed. The fire then licked up the water and dust in the trench. When the people saw what happened, they were terrified. They fell prostrate and cried, *"The LORD, he is the God! The LORD, he is the God!"* (I Kings 18:39). Elijah told the people to execute the foul prophets of Baal and not let even one of them escape. After this, Elijah said another short prayer for rain. God opened the windows of heaven and allowed rain to fall once again to quench the thirsty land.

WILLFULLY IGNORANT

At times, some of the preceding chapters may have seemed to be harsh; you may have found the tone mocking and condescending toward those who have faith and belief in evolutionism. Some may say that the tenor and gist of this book is mean-spirited. But lest we forget: Just as Elijah the prophet dealt with an insidiously evil religion, this book also confronts a wicked creed, belief, idea, and philosophy—one that has held millions in its grasp, destroyed countless souls, and harmed those for whom Christ shed His blood on the cross. As faith in Baal and worship of the goddess Asherah (Venus) deserved ridicule, devotion to evolutionism merits and has earned the right to be mocked, lampooned, and scorned. Faith, in its principles, philosophies, and ideologies, turns otherwise intelligent individuals into fools whenever they open their mouths and spew out their evolutionary ideology.

The well-known scientific law or principle, *the law of biogenesis* (also known by the Latin term *omne vivum ex vivo*), proclaims the fact that life only comes from other life, yet evolutionists have to set that God revealed, scientifically tried, true, and tested truth aside, for neither true science nor faith in God are part of evolutionary dogma. I agree with Louis Pasteur's statement *"la génération spontanée est une chimère"* ("Spontaneous generation is a dream").

As Scripture states, *"they are willingly ignorant"* (II Peter 3:5), thus, they are fools, for they ignore this fact and continue to look and search for answers that will never be forthcoming. *"For the wisdom of this world is foolishness with God. For it is written, He taketh the wise in their own craftiness"* (I Corinthians 3:19).

When I think of the phrase *"willingly ignorant,"* I cannot help but think of the recent interview of Neil deGrasse Tyson by David Freeman, Senior Science Editor for the Huffington Post. Tyson is an American astrophysicist, cosmologist, author, and science communicator. In 1997 he founded the Department of Astrophysics at the American Museum of Natural history in New York City. He is friendly, seems down to earth, has a pleasant, soft-spoken demeanor, and is a great spokesman and communicator. Below is a portion of the interview:

> DF: Do you want to talk about religion now?
> NT: I'm here for you.
> DF: Do you believe in god?
> NT: I presume you've pre-specified which god you're asking about?
> DF: Define god as you would.
> NT: You're the one who's asking the question. So pick a god and ask me if I believe in that god.
> DF: The Judeo-Christian god.
> NT: Okay, if that god is described as being all-powerful and all-knowing and all-good, I don't see evidence for it anywhere in the world. So I remain unconvinced. If that god is all-powerful and all-good, I don't see that when a tsunami kills a quarter-million or an earthquake kills a quarter-million people. I'd like to think of good as something in the interest of your health or longevity. That's a pretty simple definition of something that is good for you. That's not a controversial understanding of the word "good." So if Earth in two separate events separated by just a couple of years can kill a half-million people, then if the god as you describe exists, that god is either not all-powerful or not all-good. And so therefore I am not convinced.[1]

Although I enjoy listening to Tyson speak, he has a blind mind, for he is not convinced that there is a God. His paltry excuses do not rule out or cancel the overwhelming evidence for the existence of God. His arguments are analogous to someone looking at the space shuttle and saying, "I see no evidence of an architect or engineer." After you have read the following, see if you would agree with my assessment of Tyson:

> Reason, logic, and facts have failed to penetrate Tyson's heart, mind and conciseness, for he is a pompous evolutionist. He

proclaims what only the unenlightened, naïve, uneducated (education is the acquisition of true wisdom and true knowledge) and illiterate would grab hold of or maintain.

He can reason correctly, perceive and rightly deduce that an intelligent being created, formed, and shaped a simple flint arrowhead found lying on the ground. No amount of argumentation or reasoning will dissuade him from that rightly deduced conclusion. However, when confronted with a three-pound hunk of a squishy, pink, jelly-like organ, the most complex thing in the known universe, the human brain, in which a single cell can hold many times the amount of information as contained in any set of encyclopedias, he reasons and comes to the conclusion that it made itself.

The brain is an organ which contains approximately 100 billion neurons (nerve cells), with 100 trillion connections, and is the most mysterious, the most advanced, the most complex, sophisticated and amazing machine ever devised. It's an organ with incredibly intricate parts that is so multifarious and perplexing that even the most advanced scientific minds struggle to understand and have not even begun to plumb its depths and secrets, but somehow it made itself. And not only made itself, but came into being and now resides in the head of a creature whose ancestors sprung from the inorganic rocks that comprise much of the planet that is home to man.

He brazenly proclaims in fairytale fashion the tenets of his nefarious religious beliefs, which violate all the known laws of science and biology dealing with abiogenesis, expounding the belief that life came from non-life once upon a time.

He sees no proof or evidence that God exists. However, the proof is there, right in front of his face. Every time he stands in front of a mirror he gazes upon the world's most advanced and complex machine to ever exist. But he is determined to stay willingly ignorant, for he has too much pride and too much time and study in his delusion to honestly consider the overwhelming avalanche of evidence for God's existence.

As far as Tyson's "knowledge" of evolution, God has this to say: "Thus saith the LORD.... 'I am the LORD, that maketh all things...that turneth wise men backward, and maketh their knowledge foolish" (Isaiah 44:24–25).[2]

I thought about Tyson's words once again when I read the information in these two recent stories. As you read the following quotes, are you still unconvinced, like Tyson, that there really is no evidence that can be found in the entire world for an all-powerful and all-knowing God? Do you agree with Tyson?

> ...as fast and powerful as computers have become, they still pose no match for the human brain. Sure, a computer specifically programmed to perform singular task such as...playing chess...when we measure a computer against the entirety of what a human mind is capable of, it's not really all that close...
>
> According to biologists, the human brain has approximately 90 billion nerve cells which are linked together by, quite literally, trillions of connections called synapses. Taken together, this system of elaborate connections within the brain provides "hundreds of trillions of different pathways that brain signals travel through."
>
> ...In an effort to mimic this digitally, scientists a few years ago needed more than 82,000 processors running on one of the world's fastest supercomputers to mimic just 1 second of a normal human's brain activity...
>
> So for now computers are informational babies—they cannot "cook" for themselves.[3]

Chris Smith, writing for BGR News stated this:

> "This is a real bombshell in the field of neuroscience," Salk Institute for Biological Studies researcher Terry Sejnowski said. "Our new measurements of the brain's memory capacity increase conservative estimates by a factor of 10 to at least a petabyte, in the same ballpark as the World Wide Web." That's 1,000 terabytes, in case you were wondering...
>
> The adult brain generates about 20 watts of continuous power, as much as a dim bulb, yet it's capable of amazing things, well beyond anything any computer can currently do.[4]

There is nothing righteous, noble or redeeming about the religion of evolutionism. The pitchmen of evolution, such as Tyson, become deceivers, liars, fraudsters, and blasphemers when they write "scientific" articles for our assimilation and consumption, or when they parade their nonsense before our children in their textbooks—information that defies logic, flouts reason and science, and mocks our God, our faith, and everything sacred.

The Scripture states: "*For the invisible things of him from the creation of the world are clearly seen, being understood by the things that are made, even his eternal power and Godhead; so that they are without excuse*" (Romans 1:20). So men like Tyson who have the ability to research, examine, and investigate this earth have no defense, justification, or reason for willfully dismissing the God of Creation.

Evolutionism is a malevolent religion in which divine providence has no standing. Evolution leaves no room for creation or for any supernatural display. What is most disconcerting is the fact that so many people believe it and fail to notice the absence of the scientific method in its study. We would be well to remember that evolutionary dogma comes from the father of lies: *Satan*. This religion and its teachings do not deserve respect, honor or regard. As God told the Israelites to drive out the corrupt inhabitants of the Promised Land, we need to drive out the gospel of death, *Darwinism*, from the land with truth, answers, real science, and the "*reason of the hope that is within you...*" (I Peter 3:15).

We need to expel the unfruitful works of darkness. Research or the study of evolution wastes resources and hinders scientific advancement in all endeavors influenced and tainted by its toxins.

When Christians consider evolutionism, they need to heed Paul's words to the congregation in the city of Corinth: "*Casting down imaginations* [reasoning's] *and every high thing that exalteth itself against the knowledge of God, and bringing into captivity every thought to the obedience of Christ*" (II Corinthians 10:5). Christians must also heed Paul's instructions to the congregation in the city of Ephesus: "*And have no fellowship with the unfruitful works of darkness, but rather reprove them. For it is a shame even to speak of those things which are done of them in secret*" (Ephesians 5:11–12).

Heed also Paul's words in a letter of instruction to a young minister named Timothy: "*O Timothy! Guard what was committed to your trust, avoiding the profane and idle babblings and contradictions of what is falsely called knowledge—by professing it some have strayed concerning the faith*" (I Timothy 6:20–21 NKJV). Although Paul's directives and guidance did not specifically pertain to the discipline we call science, it was clearly referring to religious dogma, principles, laws, and tenets that were true in the spiritual realm and in the natural world, where real science is developed.

Lastly, Christians need to remember that the evidence (the fossils found in the rocks) used by evolutionists to confirm their religion is not substantiation of their faith but is firm proof of God's creative act and continuing involvement in the affairs of man. Consider the story of His judgment of unrighteous men four or five thousand years ago in a world-wide flood: "*For this they willingly are ignorant of, that by the word of God the heavens were of old, and the earth standing out of the water and in the water: Whereby the world that then was, being overflowed with water perished*" (II Peter 3:5–6).

As I close this book, I wonder if my goal, purpose, objective, and reason for writing this book will be brought to fruition. Only my readers can provide that answer. My goal was twofold. First, I sought to help strengthen and fortify the faith of Christians, whose beliefs will at times be pummeled, tested, and bashed as they amble down life's path, with truthful answers to questions about their origins. Secondly, for any others whose circle of influence may include friends of faith, but who are themselves currently outside the bounds of belief, it is my hope that you may still be curious about the things of God, which have not been drained away by the cackling of Darwin's fools. To this end, I hope this book will help to replace faith in evolutionism with faith in the God of Creation.

I have sought to present readers from both camps with logical arguments. I wanted to do as the LORD did with His people: "*Come now, and let us **reason*** [Hebrew—convince, correct, dispute, plead, and argue] *together, saith the LORD*" (Isaiah 1:18). For those who hold the Scriptures in high regard, do as Paul wrote: "*Ye that fear God, give audience*" (Acts 13:16). As Paul reasoned with the Jews through the Scriptures, I am reasoning, arguing, and making a case with those who have been born-again. As for those who are not in the encampment of faith, there are illustrations, demonstrations, and arguments available from God's creation. This was the way Paul reasoned with those who did not know the Scriptures, such as the Athenians gathered on Mars' Hill (Areopagus) in Acts 17:22–31.

In this book, I have endeavored to combine arguments from Scripture and nature with valid science and compelling prose in hopes that the reader would come to the same conclusions millions have, that "*In the beginning God created the heaven and the earth*" (Genesis 1:1), and that belief in evolution is stupid, because reindeer don't fly!

NOTES ON CHAPTER 17

1. David Freeman, "Neil deGrasse Tyson Talks God, Aliens, and Multiverses" [Interview], The Huffington Post web site, October 5, 2015, http://www.huffingtonpost.com/entry/neil-degrasse-tyson-talks-god-aliens-and multiverses_us_561297abe4b0dd85030c97fc
2. Adapted from Riemer, 2014, p. 21.

3. Yoni Heisler, "How Powerful is the Human Brain Compared to a Computer?" BGR web site, February 28, 2016, http://bgr. com/ 2016/ 02/27/power-of-the-human-brain-vs-super-computer/
4. Chris Smith, "Our Brains Might Hold 1 Petabyte of Data, which Is Almost the Entire Internet," BGR web site, January 21, 2016, http:// bgr. com/2016/01/21/brain-memory-capacity-petabyte/

What will the reader learn from the thought-provoking, comprehensive, and compelling new book *Reindeer Don't Fly* by Michael Earl Riemer? They will uncover the numerous shortcomings and misconceptions of evolutionary doctrine. But most importantly, for those who are pastors, preachers, parents, high school and college students, and all others who need answers to the excess of false evolutionary dogma paraded before them ("*science falsely so called*" I Tim. 6:20), the reader will learn the answer to questions such as:

- Scripture records that there was a world-wide flood, so where did the water come from?
- Where did the water go after the flood?
- Is there enough water to cover Mt. Everest?
- Is there a reliable method scientists can use to date the age of the earth?
- Is there life in outer space?
- Did people really co-exist with dinosaurs?
- What was the pre-Flood world really like?
- Can animals evolve; change from one kind of creature into another species?
- If the theory of evolution is invalid and worthless, why do so many scientists and others promote this teaching?

WHAT OTHERS ARE SAYING:

"The objective reader cannot ignore or dismiss the heartfelt passion, laser-like logic, and well-reasoned arguments posited throughout this book." – Freelance editor Bruce Victor Zatkow

"I've been reading your book and really enjoying it. It's refreshing to read something from a more intellectual viewpoint... Such a treasure trove of great information and logical illustrations" – Columnist Chris McMahan

"...your book! I thoroughly enjoyed it, was challenged by it, learned from it, and grew from it." – Researcher, writer, and speaker, Michelle Smallback

"...having read books similar in topic as yours and comparing them, I found *Reindeer Don't Fly* involving the reader while some of the others were very hard to read. You make the book personal with stories and experiences you had. The stories allowed the reading to not become cumbersome." – Director of Headwaters Christian Youth, Mike Prom

"I read the first 60 pages and then scanned through the rest of it. Looks really good...You dealt with the fundamental impotency of evolutionism—its inability to explain the origin of organic life...Well-documented and convincingly argued...I am definitely interested in carrying this book as one of our titles in our website bookstore." – Edward E. Stevens, International Preterist Association

www.ingramcontent.com/pod-product-compliance
Lightning Source LLC
Chambersburg PA
CBHW020348080526
44584CB00014B/941